Undergraduate Lecture Notes in Physics

Series Editors

Neil Ashby, University of Colorado, Boulder, CO, USA

William Brantley, Department of Physics, Furman University, Greenville, SC, USA

Matthew Deady, Physics Program, Bard College, Annandale-on-Hudson, NY, USA

Michael Fowler, Department of Physics, University of Virginia, Charlottesville, VA, USA

Michael Inglis, Department of Physical Sciences, SUNY Suffolk County Community College, Selden, NY, USA

Undergraduate Lecture Notes in Physics (ULNP) publishes authoritative texts covering topics throughout pure and applied physics. Each title in the series is suitable as a basis for undergraduate instruction, typically containing practice problems, worked examples, chapter summaries, and suggestions for further reading.

ULNP titles must provide at least one of the following:

- An exceptionally clear and concise treatment of a standard undergraduate subject.
- A solid undergraduate-level introduction to a graduate, advanced, or non-standard subject.
- A novel perspective or an unusual approach to teaching a subject.

ULNP especially encourages new, original, and idiosyncratic approaches to physics teaching at the undergraduate level.

The purpose of ULNP is to provide intriguing, absorbing books that will continue to be the reader's preferred reference throughout their academic career.

More information about this series at http://www.springer.com/series/8917

Reinhard Hentschke · Christian Hölbling

A Short Course in General Relativity and Cosmology

 Springer

Reinhard Hentschke
School of Mathematics
and Natural Sciences
University of Wuppertal
Wuppertal, Germany

Christian Hölbling
School of Mathematics
and Natural Sciences
University of Wuppertal
Wuppertal, Germany

ISSN 2192-4791   ISSN 2192-4805   (electronic)
Undergraduate Lecture Notes in Physics
ISBN 978-3-030-46383-0   ISBN 978-3-030-46384-7   (eBook)
https://doi.org/10.1007/978-3-030-46384-7

This Springer imprint is published by the registered company Springer Nature Switzerland AG
The registered company address is: Gewerbestrasse 11, 6330 Cham, Switzerland

Chap. 9, we focus on the development of the universe based on the thermodynamics of matter, radiation and vacuum (or cosmological constant or dark energy) as well as their equation(s) of state. Chapter 10 is devoted to the evidence we have for an accelerated expansion of the universe. The last chapter, Chap. 11, focusses on inflation, tests thereof and how we look at the Big Bang today.

There are unavoidable mathematical concepts and tools. We try to be as straightforward as possible. The path to a basic understanding of a physical concept and its experimental validation should never be longer than absolutely necessary. An exceptional example in this respect is the lectures on 'General Relativity' and 'Cosmology' by Leonard Susskind (Stanford University) available on YouTube— which we highly recommend. Their content in fact forms the inspirational backbone of this course. Textbooks we have used in preparing these notes include:

S. Weinberg *Gravitation and Cosmology.* (**1972**)
B. F. Schutz *A First Course in General Relativity.* (**1985**)
S. Weinberg *Cosmology.* (**2008**)
V. Mukhanov *Physical Foundations of Cosmology.* (**2005**)
L. D. Landau and E. M. Lifshitz *The Classical Theory of Fields.* (**1980**)
E. Poisson *A Relativist's Toolkit.* (**2004**)

A very well-written entry-level textbook on cosmology is *Introduction to Cosmology* (**2017**) by B. Ryden. Additional references are given in relation to points in the text.

We are grateful to N. Weimer for her help in creating the figures and to the students attending this course during the fall terms in 2017 and 2019 for their valuable feedback. We are also grateful to Dr. Ramon Khanna, executive editor Astronomy at Springer, for numerous fruitful discussions that helped to clarify the manuscript in a number of critical places.

e-mail: hentschk@uni-wuppertal.de hch@physik.uni-wuppertal.de

Wuppertal, Germany   Reinhard Hentschke
                                                                          Christian Hölbling

Preface

The evolution of the universe from its earliest moment to its distant future is something which fascinates both scientist and laymen alike. Within the physics curriculum, this topic is one of many, however. In addition, most physics students will choose career paths which have nothing to do with general relativity or cosmology. In other words, it will not be a disadvantage to not have taken a class in this field of science. And yet, most students choose to invest the time and energy required for at least a one-semester course. It is our goal to use this investment to lay a foundation. This foundation should enable the student to read and understand the textbooks and many of the scientific papers on the subject.

We begin this book with an overview of its content. This exposition already contains some of the important formulas, which here merely serve as markers. They will be discussed in detail in the subsequent chapters, the first of which is Chap. 2, reviewing aspects of special relativity as well as Newtonian gravitation. The next chapter, Chap. 3, is on tensor algebra, which also includes the introduction of the metric tensor itself. Chapter 4 has four sections. The first section explains the mathematical determination of intrinsic curvature via parallel transport of a vector. This will also introduce the Riemann curvature tensor. The second section is devoted to the energy-momentum tensor. In the third section, we put the two pieces together and arrive at Einstein's field equations. These equations are nonlinear in the components of the metric, which makes them hard to solve. For many applications, it is useful and permissible to work with their linearised version, the subject of the fourth section. Chapter 5 discusses the Schwarzschild solution of the field equations and classical tests of Einstein's theory. This section also includes a subsection on the detection of gravitational waves. The direct detection of gravitational waves on Earth is a recent success. But the beginnings of related experiments date back roughly 60 years, which perhaps justifies their inclusion among the 'classical' tests. Chapter 6 is devoted to black holes, including a closer look at the Schwarzschild solution as well as a discussion of the formation of black holes. Chapter 7 begins with a general overview of the history of the universe, including Hubble's law and the concept of a scale factor. Chapter 8 is devoted to Friedmann-Robertson-Walker cosmology from both the Newtonian and the general relativistic perspectives. In

Contents

Chapter 1
Overview

In this chapter we give a short overview of the topics covered in this book. We outline the path to the Einstein field equations, the Hubble law and topics beyond, all of which we will discuss in later chapters. This exposition already contains some of the important formulas, which here merely serve as markers. They will be discussed in detail in the subsequent chapters.

The central element in special relativity is the spacetime element $d\tau$, also called proper time, defined via

$$d\tau^2 = c^2\,dt^2 - dx^2 - dy^2 - dz^2 \ . \tag{1.1}$$

Note that throughout these notes, unless stated otherwise, we set the vacuum speed of light

$$\boxed{c = 1}\ , \tag{1.2}$$

e.g. the usual $c\,dt$ becomes dt. The quantity $ds^2 = dx^2 + dy^2 + dz^2$ is the infinitesimal distance between points in three-dimensional Euclidian space. The premises that $d\tau^2$ is an invariant quantity in different inertial coordinate systems[1] and that t is akin to a spatial coordinate yields the Lorentz transformations. In addition c turns out to be a limiting velocity which cannot be surpassed.

The Lagrangian $\mathcal{L}$ of a mass, m, moving with constant velocity, v, in an inertial frame can be expressed via

$$\mathcal{L} \propto d\tau/dt = \sqrt{1 - v^2}\ . \tag{1.3}$$

---

[1] In an inertial frame of reference a mass moves uniformly.

© Springer Nature Switzerland AG 2020
R. Hentschke and C. Hölbling, *A Short Course in General Relativity and Cosmology*, Undergraduate Lecture Notes in Physics,
https://doi.org/10.1007/978-3-030-46384-7_1

Fig. 1.1 Cartesian
coordinate system

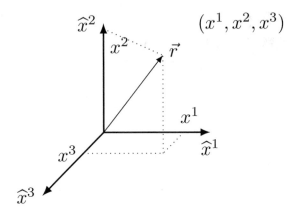

The proportionality constant follows in the limit of small v, i.e. $\mathcal{L}_{v \to 0} = const. + m\,v^2/2$. Using

$$p = \frac{\partial \mathcal{L}}{\partial v} \quad \text{and} \quad E = v\frac{\partial \mathcal{L}}{\partial v} - \mathcal{L} \tag{1.4}$$

then yields the energy-momentum relation in special relativity, i.e.

$$E^2 = p^2 + m^2 \ . \tag{1.5}$$

For $p = 0$ this is just the famous relation

$$E = m \ , \tag{1.6}$$

between energy at rest and mass.

We may wonder whether a similar approach can be applied if we do not live in an inertial frame of reference but in an accelerated one or in a gravitational field. Note that the latter two are really the only frames of reference of interest to us on a large scale.

The first question is whether it is possible to augment $d\tau$ for use in accelerated frames of reference and gravitational fields. We might expect that the general form of this new $d\tau$ or metric is given by

$$d\tau^2 = \sum_{\mu=0}^{3} \sum_{\nu=0}^{3} g_{\mu\nu} dx^\mu dx^\nu \ , \tag{1.7}$$

where $x^\mu, x^\nu = (x^0, x^1, x^2, x^3)$ with $x^0 = t, x^1 = x, x^2 = y, x^3 = z$ (cf. Fig. 1.1).[2]
Note that μ and ν are (lower or upper) indices—not powers! Note also that (1.1)
expressed in this form corresponds to

$$g_{00} = 1 \text{ and } g_{11} = g_{22} = g_{33} = -1 . \tag{1.8}$$

All other $g_{\mu\nu}$ are zero. In general, however, the $g_{\mu\nu}$ will be different and dependent
on the position in spacetime. So how do we calculate the $g_{\mu\nu}$, which are called the
components of the metric tensor?

Perhaps the problem can be simplified. Imagine a scientist working in his lab. The
lab really is a closed box with no windows. Suppose this box, which we assume is
in outer space and far enough from any gravitating mass, is subjected to a constant
acceleration of $9.81 \, \text{ms}^{-2}$ by some means. The scientist performs all kinds of experiments, but he is unable to distinguish the acceleration from the effect of gravity, i.e.
when the lab is at rest near the surface of the Earth. This is the famous 'principle
of equivalence' (of acceleration and gravitation). Note that we can also 'turn off'
gravitation by dropping the lab from some height. While the lab is in free fall, it
qualifies as inertial frame of reference and we can use (1.1) in our lab—at least for a
short while. We thus hope that if we find a solution to our problem for say accelerated
frames of reference, the same solution also applies in a gravitational field.

Note that there is a limit to the analogy between uniform acceleration and a
gravitational field. Let's again look at the freely falling laboratory. The statement was
that a scientist in the lab believes that his lab is a perfect inertial frame of reference.
However, when the laboratory is sufficiently large, the scientist will notice that the
lab is deformed by strange forces (cf. Fig. 1.2), which are called tidal forces. We, as
outside observers, immediately recognise the reason for these forces. The lab floor is
closer to the attracting mass compared to the ceiling and thus experiences an excess
force 'stretching' the lab. In addition, due to the inhomogeneity of the gravitational
field, there are inward components of the gravitational force acting on the walls. As
the lab's size increases, it becomes an increasingly poorer approximation to an inertial
frame of reference because of the tidal forces. In other words, the analogy between
the accelerated frame of reference, here the freely falling lab, and the gravitational
field only holds locally—because of the non-uniformity of the gravitational field.

'Geometry'—We digress for a moment to find out what we can learn from
geometry. Let us momentarily focus on the space-part of proper time (1.1) in two-
dimensional space, i.e. $ds^2 = dx^2 + dy^2$. Let us further assume that we are interested
in ds^2 for two nearby points on the surface of a cylinder of unit radius, i.e.

$$ds^2 = d\theta^2 + dz^2 . \tag{1.9}$$

---

[2]In order to make our expressions and equations shorter and more transparent we shall use a number
of shorthand notations. One of them is the summation convention. Using the summation convention
means that this equation can be written without the summation signs. We recognise the summations
by the fact that both μ and ν appear pairwise in the same term.

Fig. 1.2 Illustration of tidal
forces on a large laboratory
(red box) in the gravitational
fild of mass M

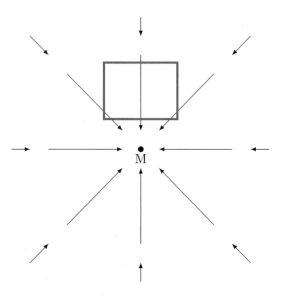

Here θ is an angular coordinate perpendicular to the long or z-direction of the cylinder.
Note that this is a flat space or Euclidian distance—just as the space-part of $d\tau$ in
special relativity. We can see this by simply cutting the cylinder along the z-direction
and smoothly flatten it out in a plane.

Now suppose we want to write down the infinitesimal distance of two points on
the surface of a unit sphere. In the next section we give the details, but here we want
to discuss the result, which is

$$ds^2 = d\theta^2 + \sin^2\theta \, d\phi^2 \, . \tag{1.10}$$

The position of a point on the sphere is uniquely described by its latitude, θ, and
longitude, ϕ. This doesn't look too terrible. It is still a simple quadratic expression in
$d\theta$ and $d\phi$. But here is the catch. Whereas (1.9) corresponds to flat space, the same
is not true for (1.10)! Why? We cannot wrap a sheet of paper around a sphere and
completely cover its surface without putting a lot of wrinkles into the paper. Note
that a cylinder's surface, even though it appears curved, is flat. This is because we
can wrap it into a sheet of paper without wrinkles (the non-destructive equivalent to
cutting). We distinguish this apparent curvature of the cylinder's surface from the real
curvature of the sphere's surface by calling the latter type of curvature 'intrinsic'.
Returning to $d\tau^2$, we conclude that it comes in two varieties—those that can be
transformed into the flat space form (1.1) and those that cannot.

Even though (1.10) cannot be transformed into a flat space ds^2 valid on the entire
surface of the sphere, locally this transformation is possible. Note that we live on the
surface of a very large sphere. But when we buy a plot of land, build a house and
finally buy flooring for its rooms, we routinely use Euclidean geometry. This works
because the Earth's surface area is so large that our daily live is rarely affected by its
intrinsic curvature.

Let's summarise: $d\tau$ in spacetime physics and ds in curved space geometry appear to be related. In physics we are able to locally get rid of a gravitational field by performing experiments in a freely falling laboratory, providing an inertial frame of reference. In an inertial frame of reference $d\tau$ is given by (1.1)—the metric of flat spacetime. In geometry a space can deviate from flatness by its intrinsic curvature. Intrinsic curvature, akin to the gravitational field of a point mass, can be removed only locally.

The scientist has an idea. What if gravity and intrinsic curvature of space are the same thing? Both can be removed locally but not globally. He talks to a mathematician about this. The mathematician tells the scientist that there is a mathematical object call the Riemann curvature tensor. The Riemann tensor can measure intrinsic curvature. And the components of this tensor are functions of the components of the metric tensor and their derivatives with respect to the coordinates. The scientist ponders this and has a second idea. If indeed gravity corresponds to curvature, and knowing that gravity is caused by the presence of mass or energy (because mass and energy are really the same thing), then why not write down an equation with the Riemann tensor on one side and mass or energy on the other side - this is a pure guess! This equation, or rather these equations, permits calculation of the $g_{\mu\nu}$.

Indeed there is a tensor in physics, called the energy-momentum tensor, which qualifies for the tensor on the right side of the field equations. The only problem is that the Riemann tensor is a tensor of rank four and the energy-momentum tensor is of rank two. However, there is a 'contraction' of the Riemann tensor, the Ricci tensor, $\mathcal{R}_{\mu\nu}$, which indeed has the same rank as the energy-momentum tensor, $T_{\nu\mu}$. After some fiddling with this the scientist finally finds the right equation

$$\mathcal{R}_{\mu\nu} - \frac{1}{2}g_{\mu\nu}\mathcal{R} = 8\pi G T_{\mu\nu} \; . \tag{1.11}$$

The quantity $\mathcal{R}$ is the Ricci scalar—'contraction' of the Ricci tensor. The $\mathcal{R}_{\mu\nu}$ as well as $\mathcal{R}$ contain the $g_{\mu\nu}$ and their derivatives as functions of the spacetime components x_μ. $T_{\mu\nu}$ on the other hand contains the sources (of gravity) like mass or energy density. Equation (1.11) are the famous Einstein field equations.

One possible solution are the sought after metric tensor components of a spherical mass distribution. We shall use them to calculate the perihelion precession of Mercury or the bending of light passing close to the Sun's surface—the two classical tests of Einstein's theory.

But how does this calculation work? Remember special relativity and the action principle, i.e.

$$\delta \int_1^2 d\tau = 0 \; . \tag{1.12}$$

The integral is over a path in spacetime connecting the spacetime points 1 and 2. δ means that we compare this path to another very similar one. The assumption is that there exists a path for which the integral will be at an extremum. If this path is compared to another one near by, then the difference is zero. But what are the variables in this case? The variables are the spacetime coordinates x^μ and their derivatives dx^μ/dp, where p is a parameterisation of the path and dp is proportional to $d\tau$. Combination of (1.7) and (1.12) yields

$$\delta \int_1^2 \sqrt{g_{\mu\nu} \frac{dx^\mu}{dp} \frac{dx^\nu}{dp}} \, dp = 0 \tag{1.13}$$

and the square root again is identified with a Lagrangian $\mathcal{L}(x^\mu, dx^\mu/dp)$. Using the Euler–Lagrange equations just like in the case of classical mechanics, where $\mathcal{L}(q, dq/dt)$ is the Lagrangian of the generalised coordinates q and velocities dq/dt, we obtain the equations of motion

$$\frac{d^2 x^\mu}{dp^2} + \Gamma^\mu_{\alpha\beta} \frac{dx^\alpha}{dp} \frac{dx^\beta}{dp} = 0 . \tag{1.14}$$

The quantities $\Gamma^\mu_{\alpha\beta} \sim \Gamma\left(g(x), dg(x)/dx\right)$ are somewhat messy functions of the $g_{\mu\nu}$ and their derivatives with respect to the spacetime components. If we know the $g_{\mu\nu}$, for instance in the vicinity of the Sun, then we can use (1.14) to calculate the path Mercury follows around the Sun or the path of a ray of light passing close to the Sun. This is just the general relativity version of Kepler's problem in classical mechanics. Only that here we shall discover that Mercury's path around the Sun is not a closed ellipse. Instead, Mercury moves on a rosetta and its perihelion rotates. The ray of light, on the other hand, is bend by the Sun. We shall do both calculations based on (1.14) and compare the results of these classical tests of general relativity to experimental data. In addition we shall study weak gravitational waves, the analogue to electromagnetic waves in the absence of sources, and discuss their detection. While in these examples general relativistic corrections are small effects, we devote the next chapter to black holes where those effects are dominant. For the simplest case of a non-rotating black hole we will see what curious new phenomena arise and we will discuss how black holes can actually form.

The aforementioned classical tests of Einstein's theory required large distances and heavy objects. But perhaps the most interesting application of general relativity is the development of the universe as a whole. We shall study a metric, which allows us to do this. The space components of this metric are multiplied by a time dependent scale factor $a(t)$, which for us will be a measure of the size of the universe at time t. The rate of change of $a(t)$ is described by

$$\frac{\dot{a}(t)}{a(t)} \equiv H(t) . \tag{1.15}$$

The quantity H is the Hubble 'constant', which is not really a constant. The 00-component of Einstein's field equations yields a differential equation relating $H(t)$ to the densities of all forms of matter and radiation. We will study the relation between these energy densities, their attendant equation of states and $H(t)$. In order to describe the experimentally observed accelerated expansion of the universe, we discuss the addition of an extra term $\Lambda g_{\mu\nu}$ to the right hand side of (1.11), where Λ is the cosmological constant.

The 3K-black body radiation, which (almost) uniformly fills today's universe, supports the idea of a universe that began very small and very hot. The possibility of a universe expanding from an originally tiny size to today's vast expanse, however, was formulated first by Georges Lemaître on the basis of certain solutions to Einstein's field equations. Later it became a necessity in the context of the synthesis of the elements. But the great uniformity of the background radiation posed a major problem for the above scenario of a Big Bang, i.e. the development of todays universe from a very small and very hot origin—as we shall discuss at some detail. A possible solution was offered by a process called cosmic inflation, stretching and thereby smoothening the universe during an early rapid expansion driven by a high 'vacuum energy'. The final stage of this process heated the universe, providing its 'classical' starting conditions.[3] Detailed measurements of the background radiation, however, showed that it is not completely smooth but exhibits thermal fluctuations of small magnitude. This in turn could be explained by quantum fluctuations, occurring during inflation, leaving their imprint on the universe after inflation had stopped and leading to the structures observable in the universe today including galaxies as well as the thermal fluctuation spectrum in the microwave background. This is not a quantum theory of gravitation, but it is a combination of both quantum theory and general relativity— similar to the famous Hawking radiation emanating from black holes. But other than the latter it has led to detectable effects in terms of the fluctuation correlations in the microwave background or the large scale correlations in the distributions of galaxies. Moreover, all these measurements, including additional independent ones, like supernova studies showing accelerated expansion of the universe or the unusually fast motion of stars in galaxies or galaxies in galaxy clusters, support the existence of so called dark matter, which is detectable only via its gravitation, and dark or vacuum energy, of which we also do not know what it is, in consistent proportions. The standard cosmological model, which is based on these ideas is called the ΛCDM model. Here Λ is the cosmological constant or vacuum energy or dark energy and CDM stands for cold dark matter. Even though the nature of its main constituents currently is completely unknown, people put a lot of trust into the model. This is because of all the different and sometimes independent experiments, which nevertheless lead to very much the same numerical values for the model parameters. In the second half of the book we shall discuss all this in simple terms, allowing the reader to develop his own confidence in the model based on hands on (approximate) calculations of (many of) the aforementioned parameter values.

---

[3] Alan Guth, the inventor of inflation, expresses this well in the title of an article, asking the question 'Was Cosmic Inflation the 'Bang' of the Big Bang? (in [1]).

Chapter 2
Review of Concepts and Some Extensions Thereof

Understanding special relativity is a prerequisite for understanding general relativity. In this chapter, we review special relativity and introduce the basic notation for describing flat spacetime in terms of general curvilinear coordinates. We also review Newtonian gravity so that it can later serve as a useful limiting case in the context of the Einstein field equations.

2.1 Special Relativity

We will be working with 'frames of reference' defined by coordinates that are established with rods and clocks. A special frame of reference is the inertial frame of reference. In an inertial frame of reference a mass moves uniformly, i.e. without acceleration, when no interactions, i.e. objects or forcefields, are present.

In the following we consider a mass moving in one dimension along the x-direction. The position of the mass is x at time t, i.e. $x(t)$. We can choose a different frame of reference, however, in which the position of the mass is $x'(t')$. Here the primed reference frame moves with the velocity w relative to the unprimed reference frame. According to our normal perception, the unprimed and the primed coordinates, should be related via the Galilei transformation[1]:

$$t' = t \quad \text{and} \quad x' = x + wt . \tag{2.1}$$

When $w > 0$ this means that the primed reference frame is moving in the negative x-direction. Note that time is the same in both coordinate frames and the velocity of the mass measured in the primed frame is simply $v' = v + w$.

---

[1]Galilei, Galileo, Italian mathematician, physicist and philosopher, *Pisa 15.2.1564, †Arcetri (today a part of Florence) 8.1.1642.

© Springer Nature Switzerland AG 2020
R. Hentschke and C. Hölbling, *A Short Course in General Relativity and Cosmology*, Undergraduate Lecture Notes in Physics,
https://doi.org/10.1007/978-3-030-46384-7_2

Our perception of time is that it has a distinct direction ('the arrow of time'), mostly due to the irreversibility of countless coupled chemical processes governing life in general. In addition, we predominantly perceive the world around us through visual information. Because the speed of light is so fast, we live under the impression that no time lag separates the actual occurrence of a distant event from our observation of this event. However, the speed of light is indeed finite and it was shown by Michelson [2] and Michelson and Morley [3] that it is independent of the inertial frame of reference. This confronts us with a problem: if we add the velocity w of the primed reference frame to the speed of light, it will not remain constant according to the Galilei transformation. Therefore the transformation needs to be revised. But how can we do this?

Instead of the special form of the transformation (2.1) we try the most unpretentious linear transformation relating the unprimed to the primed system, treating time and space on an equal footing, i.e.

$$t' = \alpha t + \beta x \quad \text{and} \quad x' = \delta t + \gamma x \tag{2.2}$$

or in differential form

$$dt' = \alpha \, dt + \beta \, dx \quad \text{and} \quad dx' = \delta \, dt + \gamma \, dx . \tag{2.3}$$

The quantities α, β, δ, and γ are thus far unknown coefficients. To calculate them we use the equation

$$\boxed{d\tau^2 = d\tau'^2} , \tag{2.4}$$

where

$$d\tau^2 \equiv dt^2 - dx^2 \quad \text{and} \quad d\tau'^2 \equiv dt'^2 - dx'^2 \tag{2.5}$$

In a box below it is shown that (2.4) is indeed true and that this is intimately tied to the constant speed of light independent of the inertial reference frame. Momentarily however we want to go ahead and use (2.4) to obtain the aforementioned coefficients. Computing $d\tau^2$ we find

$$\begin{aligned} d\tau^2 &= dt'^2 - dx'^2 \\ &= (\alpha^2 - \delta^2)dt^2 + 2(\alpha\beta - \gamma\delta)dx\,dt - (\gamma^2 - \beta^2)dx^2 \\ &= dt^2 - dx^2 . \end{aligned} \tag{2.6}$$

This implies the three conditions

$$\begin{aligned} \alpha^2 - \delta^2 &= 1 \\ \alpha\beta &= \gamma\delta \\ \gamma^2 - \beta^2 &= 1 . \end{aligned} \tag{2.7}$$

Additionally, the relative velocity of the two coordinate systems is w. Therefore we have a fourth condition

$$w = \frac{dx'}{dt'}\bigg|_{dx=0} = \frac{\delta}{\alpha} \tag{2.8}$$

The solution of this system of equations is

$$\gamma = \alpha = \frac{1}{\sqrt{1 - w^2}} \quad \text{and} \quad \beta = \delta = \frac{w}{\sqrt{1 - w^2}} . \tag{2.9}$$

Therefore the above transformation (2.2) turns into the so called Lorentz transformation[2] of special relativity, i.e.

$$\boxed{t' = \gamma(t + wx)} \quad \text{and} \quad \boxed{x' = \gamma(wt + x)} . \tag{2.10}$$

To get a feel for the new transformation we compare it to the Galilei transformation. The left sketch in Fig. 2.1 shows a rod at rest in the (x, t)-system. The color of the rod changes in regular time intervals Δt from red to blue and back. The right sketch in the same figure shows the rod in the (x', t')-system moving with a constant velocity $w > 0$ (here: $w = 0.4$) in negative x-direction. The two systems are linked via the Galilei transformation

$$\begin{pmatrix} x' \\ t' \end{pmatrix} = \begin{pmatrix} 1 & w \\ 0 & 1 \end{pmatrix} \begin{pmatrix} x \\ t \end{pmatrix} . \tag{2.11}$$

Note that $\Delta t' = \Delta t$ and that the horizontal separation of the dashed lines is the same. The latter means that the length of the rod is the same in both systems, i.e. $\Delta x = \Delta x'$.

Let's see what happens if we use the Lorentz transformation instead of the Galilei transformation, i.e.

$$\begin{pmatrix} x' \\ t' \end{pmatrix} = \gamma \begin{pmatrix} 1 & w \\ w & 1 \end{pmatrix} \begin{pmatrix} x \\ t \end{pmatrix} . \tag{2.12}$$

As seen in Fig. 2.2, the tilt of the rod indicates the coupling of space and time. Concentrating momentarily on the left endpoint of the rod we notice that the time interval Δt, the time between color changes, is different from the corresponding time interval between color changes of the same endpoint in the primed system, $\Delta t'$. In addition, the horizontal distances between the dashed lines, Δx and $\Delta x'$, tracing the paths of the endpoints is different as well. This is shown more clearly in Fig. 2.3.

---

[2]Lorentz, Hendrik Antoon, Dutch physicist, *Arnheim 18.7.1853, †Haarlem 4.2.1928; Nobel Prize in physics 1902 together with P. Zeeman.

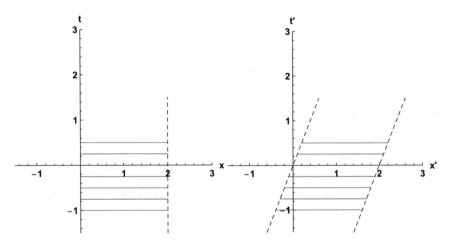

Fig. 2.1 Left: A rod changing color from blue to red and back in regular time intervals at rest in the (x, t)-system. Right: The same rod in the moving system according to the Galilei transformation. Dashed lines trace the endpoints of the rod in their respective system

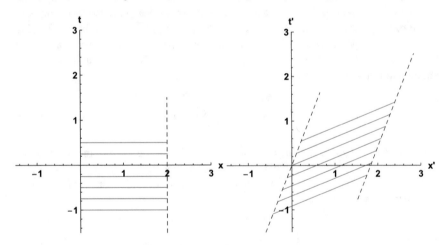

Fig. 2.2 Left: A rod changing color from blue to red and back in regular time intervals at rest in the (x, t)-system. Right: The same rod in the moving system according to the Lorentz transformation

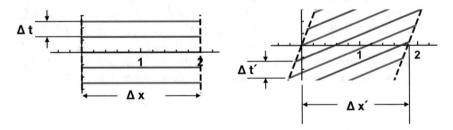

Fig. 2.3 The meaning of Δt, $\Delta t'$, Δx and $\Delta x'$ in a closeup view of the previous figure

Calculating the relation between Δt, $\Delta t'$, Δx and $\Delta x'$ we find

$$\boxed{\Delta t' = \gamma \Delta t} \quad \text{and} \quad \boxed{\Delta x' = \frac{1}{\gamma}\Delta x}. \tag{2.13}$$

The equation for $\Delta t'$ follows by inserting two spacetime points of the left endpoint of the rod in the rest system, with a time separation of Δt, into the Lorentz transformation. The other relation we find by calculating the intersections of the dashed lines with the x'-axis in the right panel of Fig. 2.3. The first relation is the mathematical formulation of the somewhat imprecise statement that clocks slow down in a moving system. The second relation means that rods are shorter (in the direction of motion) from the perspective of a moving observer. Note however that in between the aforementioned intersections the color changes along the x'-axis, indicating that the respective coordinates correspond to different times in the unprimed system.

> • **Example—Time Dilatation and Muon Decay:** Interaction of cosmic radiation with the Earth's atmosphere leads to the creation of muons at a height of about 20 km above the surface of the Earth. These muons possess a mean lifetime of $t_M \approx 2.2\ \mu s$ in their rest frame. Their kinetic energy translates into $\gamma \approx 50$. Thus, according to (2.13), from the point of view of an observer at the origin of a coordinate frame on the surface of the Earth, t_M is stretched to γt_M. This means that instead of only $\approx c\, t_M \approx 600$ m (note the explicit use of c), the muons are able to survive for $\approx c\, \gamma\, t_M \approx 33$ km and therefore they can be detected on Earth's surface.

The red line in the left panel of Fig. 2.4 is the world line of a signal traveling at the speed of one (in our units) in the (x, t)-system, i.e. its angle width respect to the x-axis is 45°. Converting this line to the (x', t')-system yields exactly the same 45°-line. Below we show that one, in our units, indeed is a limiting velocity—which in real live is the vacuum speed of light.[3] The dashed lines in Fig. 2.4 are the world line of an observer at rest in the primed system, which itself is moving at $w = 0.4$ along the negative x-direction. Note that the observer's world line intersects with the light or photon world line at different coordinates in the two systems. Having said this we are now ready to resume our mathematical discussion of the Lorentz transformation.

> • **Derivation of** (2.4): The left sketch in the following figure shows two light beams simultaneously emitted by a source at rest in the unprimed system whose location is indicated by the lower red dot. The beams, traveling in

---

[3]This connection is made in electrodynamics.

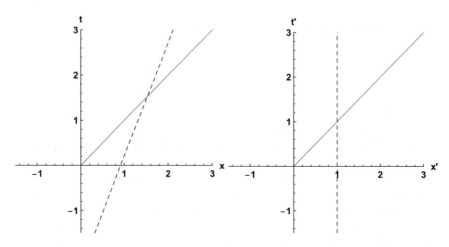

Fig. 2.4 Left: World line of a photon emitted from a source in the unprimed system (red solid line) and the world line of an observer at rest in the primed system (black dashed line). Right: The two world lines in the primed system moving with speed $w = 0.4$ along the negative x-direction

opposite directions, are reflected by two mirrors (vertical gray lines) at identical distances from the source and meet again in the upper red dot. The right sketch depicts the same sequence of events from within a moving reference frame, where the shaded area is tilted and distorted. However, the independence of the speed of light from the inertial reference frame implies that all arrows are at $45°$ to the coordinate axes in their respective reference frame. Moreover, the shaded area A in the left sketch contains spacetime events which are all mapped into the shaded area A' of the right sketch and vice versa! This implies $A = A'$.

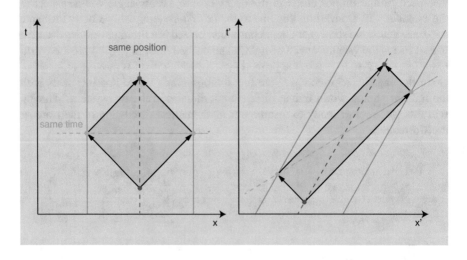

In combination the 45° angles of the light beams together with $A = A'$ imply the validity of the following sketch

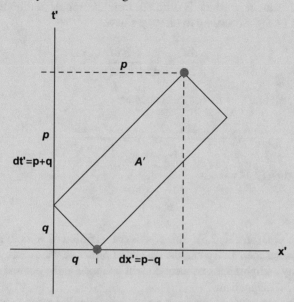

Note that $A' = 2pq$, $dt' = p + q$ and $dx' = p - q$. Here dt' is the time separation of the two events indicated by the red dots, i.e. emission and absorption of the light and dx' is the corresponding spatial separation. The same holds true in the unprimed system, where however $dx = 0$. Thus we find $2A' = dt'^2 - dx'^2 = dt^2 - dx^2 = 2A$ - proving (2.4).

At first glance all this may seem somewhat special. However, we can start with two red dots connected by a red dashed line—as in the sketch. The red dashed line is the world line of an arbitrary mass at rest or moving with a speed lower than the speed of light between two arbitrary spacetime points separated by dt and dx or dt' and dx'. Subsequently we add in the light beams and mirrors and arrive at the same result. So this is quite general.

The world line of an arbitrary mass will always possess an angle less than 45° with respect to the time axis. It connects timelike events. As the angle approaches 45° the above rectangle narrows and stretches—approaching an infinite straight line. However, for every finite $|dx| = |dt|$ the proper time $d\tau = 0$ along this 45°-line, which is called lightlike. Any two events connected by a line possessing an angle greater than 45° with respect to the time axis are called spacelike events. Because $d\tau^2$ is the same in every inertial frame, the type of separation between two events (timelike, lightlike or spacelike) has an invariant meaning and cannot be changed by going to a different inertial frame.

As already mentioned, in our units $w = 1$ is a limiting velocity, which no mass can surpass! For instance, γ for $w > 1$ becomes complex. In particular, there is the addition of velocities, which is different from the simple addition we had used previously (cf. (2.1)). According to (2.10) we have

$$\frac{\mathrm{d}x'}{\mathrm{d}t'} = \frac{\mathrm{d}x + w\mathrm{d}t}{\mathrm{d}t + w\mathrm{d}x} \ .$$

With $v' = \mathrm{d}x'/\mathrm{d}t'$ and $v = \mathrm{d}x/\mathrm{d}t$ this becomes

$$v' = \frac{v + w}{1 + wv} \quad \text{or} \quad v = \frac{v' - w}{1 - wv'} \ . \tag{2.14}$$

If now $v = 1$, then (2.14) yields

$$v' = 1 \ , \tag{2.15}$$

even though $w \neq 0$ (cf. Fig. 2.5). This means that addition of velocities cannot lead to a velocity larger than 1. We emphasise that in principle we do not yet know that 1 is the velocity of light! This connection will be made in the context of the theory of electricity and magnetism.

At this point we return to the least action principle. We want to write down an action for a relativistic free particle moving in two-dimensional spacetime. The only quantity, which we know thus far, that has something to do with a path is $\mathrm{d}\tau$. So let's try

$$S \propto -\int_1^2 \mathrm{d}\tau = -\int_{t_1}^{t_2} \frac{\mathrm{d}\tau}{\mathrm{d}t} \mathrm{d}t = -\int_{t_1}^{t_2} \sqrt{1 - v^2} \mathrm{d}t \ , \tag{2.16}$$

where 1 and 2 are two points in spacetime and $v = \mathrm{d}x/\mathrm{d}t$. The minus sign, at this point, is just a convention. Let's work out $\delta S = S(v + \delta v) - S(v)$, i.e.

$$\delta S \propto \int_{t_1}^{t_2} \mathrm{d}t \, v \frac{\delta v}{\sqrt{1-v^2}} \tag{2.17}$$
$$+ \int_{t_1}^{t_2} \mathrm{d}t \, \frac{1}{2} \frac{\delta v^2}{\left(1-v^2\right)^{3/2}} + \mathcal{O}(\delta v^3) \ .$$

We may write $\delta v = \mathrm{d}\delta x/\mathrm{d}t$ and then use partial integration to rewrite the first integral, i.e.

$$\delta S \propto -\int_{t_1}^{t_2} \mathrm{d}t \, \delta x \frac{\mathrm{d}}{\mathrm{d}t} v \frac{1}{\sqrt{1-v^2}} \tag{2.18}$$
$$+ \int_{t_1}^{t_2} \mathrm{d}t \, \frac{1}{2} \frac{\delta v^2}{\left(1-v^2\right)^{3/2}} + \mathcal{O}(\delta v^3) \ .$$

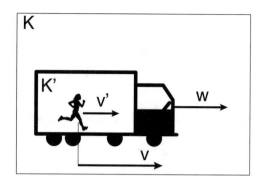

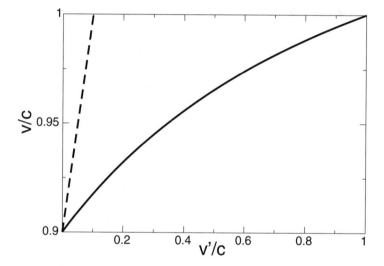

Fig. 2.5 Top: Reference frame K' is moving relative to reference frame K with the velocity w. In this figure we use the letter c explicitly. A runner moves with velocity v' relative to the origin of K' in the same direction. Relative to the origin of K the runner's velocity is v. Bottom: v plotted vs. v' according to (2.14) when $w = 0.9c$. The dashed straight line is the naive (non-relativistic) addition of velocities, $v = v' + w$. Deviation from the relativistic result increases when w approaches c

Because δx is arbitrary,[4] the first integral vanishes only if the integrand vanishes, i.e.

$$\frac{d}{dt} v \frac{1}{\sqrt{1-v^2}} = \left(1-v^2\right)^{-3/2} \frac{dv}{dt} = 0 . \tag{2.19}$$

This means that v must be constant, i.e. the path of a free particle in our two-dimensional spacetime, for which the action has an extremum, is a straight line connecting the points 1 and 2. In addition, the second integral is positive ($v < 1$!),

---

[4]The only requirement is that the velocity along an alternative path is less then 1.

which means that the extremum is a minimum of S, provided that the a yet unspecified proportionality constant is positive.

According to (2.16) the Lagrangian in the present case, up to a constant, is

$$\mathcal{L} \propto -\sqrt{1 - v^2} \ .$$

The proportionality constant follows if we consider the limit of small velocity, i.e. $v \ll 1$, via

$$\mathcal{L} \propto -\sqrt{1 - v^2} \approx -\left(1 - \frac{1}{2}v^2\right) \ .$$

In order for the second term to be equal to $\frac{1}{2}mv^2$, the proportionality constant must be m, i.e.

$$\mathcal{L} = -m\sqrt{1 - v^2} \ . \tag{2.20}$$

First we calculate the relativistic momentum of the point mass, i.e.

$$p = \frac{\partial \mathcal{L}}{\partial v} = m\gamma v \ . \tag{2.21}$$

The factor $m\gamma$ is also known as the relativistic mass. The latter approaches infinity as $v \to 1$! A second interesting quantity is the energy given by

$$E = v\frac{\partial \mathcal{L}}{\partial v} - \mathcal{L} = m\gamma v^2 + m\gamma^{-1} = m\gamma \ . \tag{2.22}$$

Another way to express this is

$$\boxed{E^2 = p^2 + m^2} \ . \tag{2.23}$$

In particular for $v \to 0$ we obtain the famous formula

$$\boxed{E = m} \tag{2.24}$$

for the rest energy .

• **Lorentz transformation as 'rotation':** The figure shows a vector relative to two coordinate systems. The primed system is rotated counterclockwise by the angle φ relative to the unprimed system. We have

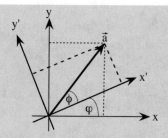

$$\begin{pmatrix} x' \\ y' \end{pmatrix} = \begin{pmatrix} \cos\varphi & \sin\varphi \\ -\sin\varphi & \cos\varphi \end{pmatrix} \cdot \begin{pmatrix} x \\ y \end{pmatrix} . \tag{2.25}$$

Inserting $x = r\cos\theta$ and $y = r\sin\theta$, i.e. polar coordinates, yields

$$r' \begin{pmatrix} \cos\theta' \\ \sin\theta' \end{pmatrix} = r \begin{pmatrix} \cos\varphi\cos\theta + \sin\varphi\sin\theta \\ -\sin\varphi\cos\theta + \cos\varphi\sin\theta \end{pmatrix}$$

$$= r \begin{pmatrix} \cos(\theta - \varphi) \\ \sin(\theta - \varphi) \end{pmatrix} . \tag{2.26}$$

Thus $r' = r$ and $\theta' = \theta - \varphi(= \phi)$.

Now we repeat this for the Lorentz transformation (2.12). Instead of $\sin\alpha$ and $\cos\alpha$, however, we use the hyperbolic functions

$$\sinh\alpha = \frac{1}{2}\left(e^\alpha - e^{-\alpha}\right)$$

and $\tag{2.27}$

$$\cosh\alpha = \frac{1}{2}\left(e^\alpha + e^{-\alpha}\right) ,$$

which satisfy the identity $\cosh^2\alpha - \sinh^2\alpha = 1$. Now we define

$$x = \rho\cosh\alpha \quad \text{and} \quad t = \rho\sinh\alpha . \tag{2.28}$$

Thus

$$\rho' \begin{pmatrix} \cosh\alpha' \\ \sinh\alpha' \end{pmatrix} = \begin{pmatrix} \cosh\omega & \sinh\omega \\ \sinh\omega & \cosh\omega \end{pmatrix} \cdot \rho \begin{pmatrix} \cosh\alpha \\ \sinh\alpha \end{pmatrix} , \tag{2.29}$$

where $\cosh\omega \equiv \gamma$ and $\sinh\omega \equiv \gamma w$. Note that $\cosh^2\omega - \sinh^2\omega = \gamma^2 - \gamma^2 w^2 = \gamma^2(1 - w^2) = 1$. Finally,

$$\rho' \begin{pmatrix} \cosh \alpha' \\ \sinh \alpha' \end{pmatrix} = \rho \begin{pmatrix} \cosh \omega \cosh \alpha + \sinh \omega \sinh \alpha \\ \sinh \omega \cosh \alpha + \cosh \omega \sinh \alpha \end{pmatrix} \tag{2.30}$$

$$= \rho \begin{pmatrix} \cosh(\omega + \alpha) \\ \sinh(\omega + \alpha) \end{pmatrix} .$$

Thus $\rho' = \rho$ and $\alpha' = \omega + \alpha$.

Throughout the section we have always used the x-axis as a proxy for an arbitrary spatial direction. It is obvious though that there is nothing special about this choice of coordinates and we could have used any other direction. The momentum square in (2.23) obviously generalizes to the momentum vector square and the general form of proper time is

$$d\tau^2 = dt^2 - dx^2 - dy^2 - dz^2 = dx^\mu dx_\mu . \tag{2.31}$$

2.2   Newtonian Gravity

In Newtonian physics

$$\vec{F} = m\vec{a} , \tag{2.32}$$

where $\vec{F}$ is the force on the mass m giving rise to the acceleration $\vec{a}$. This is a three-vector equation. For every component $v^i = dx^i/dt$ is the velocity at time t and $a^i = d^2x^i/dt^2$.[5]

In Galileian gravity the surface of the Earth is flat and the gravitational force

$$\vec{F}_g = m\vec{g} \quad (g \approx 9.8 \text{ ms}^{-2}) \tag{2.33}$$

does not depend on how high you are above that surface. Combination of (2.32) and (2.33) yields

$$d^2\vec{r}/dt^2 = \vec{g} . \tag{2.34}$$

The masses cancel and the motion does not depend on the mass of the object. This is the simplest version of the 'equivalence principle'. If we let a gas cloud fall, then the shape of that cloud does not change. We assume of course that the shape of the cloud does not change due to the relative motion of the particles to each other. The gravitational field is therefore undetectable within the cloud. Below we have to modify this, because of 'tidal forces'. Tidal forces occur because the Earth is not flat but curved. Therefore the gravitational force depends on position.

------

[5]In the remainder of this subsection upstairs numbers are exponents and not indices!.

Now let's move on from flat space-gravity to Newton's gravity, i.e.

$$F_g = G\frac{mM}{r^2} \quad (G \approx 6.7 \cdot 10^{-11} \text{ m}^3\text{kg}^{-1}\text{s}^{-2}) .$$ (2.35)

Note that the gravitation constant G is required because of dimensional consistency. The force on either mass, m or M, separated by the distance r is oriented towards the other mass.

Gravity is very weak! What we notice is the attraction between objects and Earth. But we do not observe attraction between objects of any size. On the other hand, attraction between things is common (magnets, adhesive tape, ...). Nevertheless, at very large distances gravity is the only force that matters. It is attractive only and its range is infinite:

$$\lim_{R\to\infty} \int^R dr\, r^{d-1} r^{-n} = \begin{cases} \infty & (n \le d) \\ \text{finite} & (n > d) \end{cases}$$ (2.36)

(d: space dimension).

How did Newton get the inverse-square-law? Observation! According to Kepler's third law $T \propto R^{3/2}$, where T is the time for a planet to complete a full revolution around the Sun and R is the semi-major axis of its ellipse. If for simplicity we assume that the ellipse is a circle, then the circular acceleration must be balanced by gravity, i.e. $\omega^2 R \propto R^{-2}$ and with $\omega = 2\pi/T$ we confirm the above law.

Let us study Newton's law of gravitation for a while.[6] If m is the Earth's mass and M is the mass of the Sun, how can we account for the gravitational effects of the Moon and the other large bodies in the solar system and possibly beyond? The mathematical answer is

$$m\ddot{\vec{r}} = Gm \sum_j \frac{M_j}{|\vec{r}_j - \vec{r}|^2} \frac{\vec{r}_j - \vec{r}}{|\vec{r}_j - \vec{r}|} .$$ (2.37)

Here $\vec{r}$ is the position vector of the Earth in a suitable coordinate system. The vector $\vec{r}_j - \vec{r}$ joins the Earth with the (celestial) body j possessing the mass M_j. This states that gravitation is simply additive.

There is a conceptual problem here as you may have noticed. What do we mean by separation of two masses? Do we mean the distance between their centers or do we mean the distance between their surfaces or some other distance?

Let's pretend Earth is the size of a pinhead with a diameter of 3 mm. Then the Sun is a basketball some 35 m away. Here the precise meaning of distance between the two masses may not be very important. The situation changes completely if we consider a satellite in an orbit near the Earth's surface. Even more difficult is the situation if we consider the motion of a pendulum directly on the Earth's surface. What is the proper mass-to-mass separation in these cases? One can show (cf. below)

---

[6]The following is adopted from [4].

that two radially symmetric mass distributions, possessing the total masses m and M, each feel attracted to the other according to the law (2.35). The proper r is the distance separating the midpoints of the two mass distributions.[7]

We want to show that the last statement is true. Suppose a large mass is cut up into many volume elements each contributing a small increment δm_j to the total mass. Thus each volume element (or mass element) contributes to the total gravitational force $\vec{F}_g(\vec{r})$ at position $\vec{r}$ given by

$$\vec{F}_g(\vec{r}) = Gm \sum_j \frac{\delta m_j}{|\vec{r}_j - \vec{r}|^2} \frac{\vec{r}_j - \vec{r}}{|\vec{r}_j - \vec{r}|} \qquad (2.38)$$

(cf. (2.37)). Momentarily we assume that we measure this force via a point mass m located at position $\vec{r}$. A point mass is a mathematical approximation, which assumes that the entire mass is concentrated in a point.

The sum is inconvenient to deal with. Therefore we decide to convert it into an integral, i.e.

$$\sum_j \delta m_j = \int_{V_b} \rho(\vec{r}') \mathrm{d}V' \ .$$

Here V_b is the total volume of the large mass or body, and $\rho(\vec{r}')$ is the mass density or mass distribution inside the large body at the position $\vec{r}'$. Note that the two sides of this equation indeed are equal, because we obtain the total mass in both cases. On the right side we of course assume that the mass increments can be arbitrarily small, i.e. $\lim_{\delta m_j \to 0} \delta m_j / \delta V = \rho(\vec{r}')$, where $\vec{r}'$ is the position of the small volume δV containing δm_j.

Thus we have

$$\vec{F}_g(\vec{r}) = Gm \int_{V_b} \mathrm{d}V' \frac{\rho(\vec{r}')}{|\vec{r}' - \vec{r}|^2} \frac{\vec{r}' - \vec{r}}{|\vec{r}' - \vec{r}|} \ . \qquad (2.39)$$

In general this integral is difficult to solve. However, if the distribution $\rho(\vec{r}')$ is radially symmetric, i.e. $\rho(\vec{r}') = \rho(r')$, then we can achieve great simplification.

First we note that

$$\vec{\nabla}_{\vec{r}} \cdot \vec{F}_g = 0 \qquad (2.40)$$

outside V_b. We apply Gauss' theorem or the divergence theorem to this equation, i.e.

$$0 = \int_V \vec{\nabla}_{\vec{r}} \cdot \vec{F}_g \mathrm{d}V = \oint_{\partial V} \vec{F}_g \cdot \mathrm{d}\vec{f} \ . \qquad (2.41)$$

---

[7]In examples involving the Earth's gravitation, we shall always approximate Earth as a radially symmetric mass distribution.

Fig. 2.6 Spherical mass
distribution at the center
(black circle) and attendant
integration volume (grey
shading)

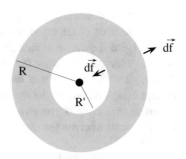

The integration volume, V, indicated by the darker shaded area in Fig. 2.6, is the
volume between two concentric spherical shells. Both shells as well as the volume
V_b, shown as the small black circle, are centred on the same origin. Each of the
two shells completely includes V_b. From the radial symmetry of the problem it is
clear that m experiences a force pulling it straight towards the origin, i.e. the center
of the mass distribution, regardless of where m is located (momentarily outside
V_b). This means that the magnitude of $\vec{F}_g$ is constant on each of the two shells, i.e.
$F_g = F_g(R)$ on the outer shell with radius R and $F_g = F_g(R')$ on the inner shell with
radius R'. In addition, on the outer shell $\vec{F}_g \cdot d\vec{f} < 0$, because according to Gauss'
theorem $d\vec{f}$ must point perpendicularly away from the integration volume's surface.
Thus $\vec{F}_g \cdot d\vec{f} > 0$ on the inner shell's surface, because there $\vec{F}_g$ and $d\vec{f}$ are parallel.
Equation (2.41) therefore yields

$$0 = -4\pi R^2 F_g(R) + 4\pi R'^2 F_g(R') . \tag{2.42}$$

So far we have only stated that $R > R'$ and that both radii are larger than the radius
of the mass distribution. Otherwise we may choose arbitrary and independent values
for R and R'. From this and (2.42) we conclude

$$R^2 F_g(R) = \text{constant} . \tag{2.43}$$

We determine the constant by considering the limit $R \to \infty$. Note that in this limit
the mass distribution reduces to a point mass and thus

$$\vec{F}_g(\vec{r}) = -\frac{GmM}{r^2} \frac{\vec{r}}{r} , \tag{2.44}$$

where $M = \int_{V_b} \rho(r') dV'$. This means that a radially symmetric mass distribution,
$\rho(r')$, on its outside, gives rise to the same gravitational field as a point mass M
(Here by gravitational field we mean the right side of (2.44) without the point mass
m.). By extension of this reasoning we conclude that two radially symmetric mass
distributions, which do not overlap, possess the same gravitational interaction as their
corresponding point masses located at the center of the respective mass distribution.

Before we continue, we want to discuss the gravitational force inside a radially symmetric mass distribution. First we remove the black circle, i.e. the mass distribution, from the above sketch and insert instead a thin spherical shell, uniformly covered with mass, between the two spherical shells defining the integration volume. Again, all shells are centred on the same origin. Equation (2.41) still does apply. Except now, because of the symmetry of the problem, $F_g(0) = 0$. From (2.42) we conclude that $F_g = 0$ everywhere inside the mass-covered shell. This is Newton's first theorem, whereas the statement following (2.44) is known as Newton's second theorem.

If we measure the force of gravitation 'tunnelling' through a radially symmetric mass distribution, $\rho(r)$, with a point mass along a straight line through the distribution's center, what do we find? Outside the mass distribution we measure $F_g(r) \propto M r^{-2}$, where r is the distance to the center of $\rho(r)$ and M is the distribution's total mass. Inside the mass distribution we measure $F_g(r) \propto M(r) r^{-2}$ instead, where $M(r)$ is the mass enclosed in a sphere with radius r. As we have just shown, there is no contribution from beyond r. If the mass distribution is uniform, i.e. $\rho(r) = \rho$, then $M(r) = 4\pi \rho r^3 / 3$ and in this case $F_g(r) \propto \rho r$. Because the force inside the mass distribution must continuously tie on to the force on the outside given by (2.44), we find

$$\vec{F}_g(\vec{r}) = -\frac{GmM}{R^3} r \frac{\vec{r}}{r} \, , \tag{2.45}$$

when $r \leq R$.

Remark Equations (2.42) and (2.43) may convey the impression that they 'prove' the r^{-2}-dependence of the gravitational force. This is not true. The r^{-2}-dependence already enters through (2.40)!

Finally we want to introduce the gravitational potential $\varphi(\vec{r})$ via

$$\varphi(\vec{r}) = -\frac{GM}{r} \, . \tag{2.46}$$

Note that $-\vec{\nabla}_{\vec{r}} \, \varphi(\vec{r}) = \vec{F}_g / m = \vec{a}$ according to (2.44). Moreover (2.40) yields

$$\Delta_{\vec{r}} \, \varphi(\vec{r}) = 0 \, . \tag{2.47}$$

in the region where there are no sources (or masses). In analogy to the Poisson equation from electrodynamics we may extend this to

$$\Delta_{\vec{r}} \, \varphi(\vec{r}) = 4\pi G \rho(\vec{r}) \, , \tag{2.48}$$

where ρ is an appropriate source term. Notice a certain similarity to the field equations (1.11), which we want to develop here.

2.3 Generalised Equations of Motion

This section uses notation that will be explained in detail in the following chapter. But we want to get at least some feeling for the physical side of relativity before concentrating on its mathematical foundation.

2.3.1 Generalised Equations of Motion via Equivalence Principle

Consider a particle gently released by the scientist inside the freely falling lab. One of the lab's corners provides the axes of a coordinate system whose coordinates are ξ^α. The particle moves on a straight line in spacetime, i.e.

$$\frac{d^2\xi^\alpha}{d\tau^2} = 0 . \tag{2.49}$$

Now take the perspective of an observer watching the particle from his own arbitrary coordinate system with coordinates $x^\mu = x^\mu(\xi^\alpha)$ so that the free falling coordinates can be expressed as $\xi^\alpha = \xi^\alpha(x^\mu)$. Hence

$$0 = \frac{d}{d\tau}\frac{\partial\xi^\alpha}{\partial x^\mu}\frac{dx^\mu}{d\tau} = \frac{\partial\xi^\alpha}{\partial x^\mu}\frac{d^2x^\mu}{d\tau^2} + \frac{\partial^2\xi^\alpha}{\partial x^\mu\partial x^\nu}\frac{dx^\mu}{d\tau}\frac{dx^\nu}{d\tau} .$$

Multiplying this by $\partial x^\lambda/\partial\xi^\alpha$ and using

$$\frac{\partial x^\lambda}{\partial\xi^\alpha}\frac{\partial\xi^\alpha}{\partial x^\mu} = \delta^\lambda_\mu \tag{2.50}$$

yields the general equation of motion

$$0 = \frac{d^2x^\lambda}{d\tau^2} + \Gamma^\lambda_{\mu\nu}\frac{dx^\mu}{d\tau}\frac{dx^\nu}{d\tau} . \tag{2.51}$$

The quantities

$$\Gamma^\lambda_{\mu\nu} = \frac{\partial x^\lambda}{\partial\xi^\alpha}\frac{\partial^2\xi^\alpha}{\partial x^\mu\partial x^\nu} \tag{2.52}$$

are called Christoffel symbols or affine connection.

2.3.2 Generalised Equations of Motion via Least Action Principle

Let's derive the same equation of motion using the least action principle. Just as in special relativity we start from

$$\delta \int d\tau = 0 . \tag{2.53}$$

Here however

$$d\tau = \sqrt{g_{\alpha\beta} dx^\alpha dx^\beta} \tag{2.54}$$

is a generalisation of (2.31) and we assume that the $g_{\alpha\beta} = g_{\beta\alpha}$ are functions of the x^α.[8] Hence

$$\delta \int \sqrt{g_{\alpha\beta} \frac{dx^\alpha}{d\tau} \frac{dx^\beta}{d\tau}} d\tau = 0 . \tag{2.55}$$

Note that the Lagrangian is $\mathcal{L}(x, dx/d\tau) = \sqrt{g_{\alpha\beta} \frac{dx^\alpha}{d\tau} \frac{dx^\beta}{d\tau}}$. Using the Euler–Lagrange equations, i.e.

$$\frac{d}{d\tau} \frac{\partial \mathcal{L}}{\partial (\partial x^\gamma / \partial \tau)} - \frac{\partial \mathcal{L}}{\partial x^\gamma} = 0 , \tag{2.56}$$

we have

$$\frac{d}{d\tau} \left(\frac{1}{\sqrt{\cdots}} g_{\gamma\alpha} \frac{dx^\alpha}{d\tau} \right) - \frac{1}{2} \frac{1}{\sqrt{\cdots}} \frac{\partial g_{\alpha\beta}}{\partial x^\gamma} \frac{dx^\alpha}{d\tau} \frac{dx^\beta}{d\tau} = 0 .$$

With $\sqrt{\cdots} = 1$ this becomes

$$\frac{\partial g_{\gamma\alpha}}{\partial x^\beta} \frac{dx^\alpha}{d\tau} \frac{dx^\beta}{d\tau} + g_{\gamma\alpha} \frac{d^2 x^\alpha}{d\tau^2} - \frac{1}{2} \frac{\partial g_{\alpha\beta}}{\partial x^\gamma} \frac{dx^\alpha}{d\tau} \frac{dx^\beta}{d\tau} = 0$$

or, using the obvious symmetry of $g_{\alpha\beta}$,

$$g_{\gamma\alpha} \frac{d^2 x^\alpha}{d\tau^2} + \frac{1}{2} \left(\frac{\partial g_{\gamma\alpha}}{\partial x^\beta} + \frac{\partial g_{\beta\gamma}}{\partial x^\alpha} - \frac{\partial g_{\alpha\beta}}{\partial x^\gamma} \right) \frac{dx^\alpha}{d\tau} \frac{dx^\beta}{d\tau} = 0 .$$

---

[8]The reason for putting indices downstairs instead of upstairs will be explained in the next section. Here we merely adhere to the rule that summation over paired indices requires one index to be downstairs (e.g. $_\alpha$) and the other one to be upstairs (e.g. $^\alpha$).

The final step is the multiplication with the inverse of **g** defined via

$$g^{\mu\gamma} g_{\gamma\alpha} = \delta^{\mu}_{\alpha} , \tag{2.57}$$

i.e.

$$\boxed{\frac{\mathrm{d}^2 x^{\mu}}{\mathrm{d}\tau^2} + \Gamma^{\mu}_{\alpha\beta} \frac{\mathrm{d}x^{\alpha}}{\mathrm{d}\tau} \frac{\mathrm{d}x^{\beta}}{\mathrm{d}\tau} = 0} \tag{2.58}$$

with

$$\boxed{\Gamma^{\mu}_{\alpha\beta} = \frac{1}{2} g^{\mu\gamma} \left(\frac{\partial g_{\alpha\gamma}}{\partial x^{\beta}} + \frac{\partial g_{\gamma\beta}}{\partial x^{\alpha}} - \frac{\partial g_{\alpha\beta}}{\partial x^{\gamma}} \right)} . \tag{2.59}$$

The last equation simply follows from the comparison with the previous result in (2.51). Note that (2.59) is general. Equation (2.52) is a special case, because it contains the coordinates ξ^{μ} of the inertial frame of reference. The equation of motion (2.58) is also referred to as the geodesic equation and the corresponding paths of minimal action are the geodesics. Note that from (2.59) we see that the Christoffel symbols are symmetric in the lower indices

$$\Gamma^{\mu}_{\alpha\beta} = \Gamma^{\mu}_{\beta\alpha} \tag{2.60}$$

Let's make contact with Newtonian gravity. We assume that the gravitational field is 'small' and that all bodies move much slower than the speed of light. Then

$$g_{\alpha\beta} = \begin{pmatrix} +1 & 0 & 0 & 0 \\ 0 & -1 & 0 & 0 \\ 0 & 0 & -1 & 0 \\ 0 & 0 & 0 & -1 \end{pmatrix} + \text{'small'}_{\alpha\beta} \tag{2.61}$$

and

$$g^{\alpha\beta} = \begin{pmatrix} +1 & 0 & 0 & 0 \\ 0 & -1 & 0 & 0 \\ 0 & 0 & -1 & 0 \\ 0 & 0 & 0 & -1 \end{pmatrix} + \text{'small'}^{\alpha\beta} \tag{2.62}$$

Obviously (2.57) is satisfied by the leading terms. Because $v \ll 1$ we can write $\mathrm{d}\tau \approx \mathrm{d}t$. Setting $\alpha = \beta = 0$ and $\mu = 1, 2$ or 3 in (2.58) yields

$$\frac{\mathrm{d}^2 x^{\mu}}{\mathrm{d}t^2} \approx -\Gamma^{\mu}_{00} \tag{2.63}$$

or with (2.59)

$$\frac{d^2x^\mu}{dt^2} \approx -\frac{1}{2}g^{\mu\gamma}\Big(\underbrace{\frac{\partial g_{0\gamma}}{\partial x^0}}_{=0} + \underbrace{\frac{\partial g_{\gamma0}}{\partial x^0}}_{=0} - \frac{\partial g_{00}}{\partial x^\gamma}\Big)$$

$$= \frac{1}{2}g^{\mu\gamma}\frac{\partial g_{00}}{\partial x^\gamma}$$

$$= -\frac{1}{2}\frac{\partial g_{00}}{\partial x^\mu} . \tag{2.64}$$

Here we use the static limit, i.e. the $g_{\gamma0}$ do not depend on time. In Newton's theory

$$\frac{d^2x^\mu}{dt^2} = -\frac{\partial}{\partial x^\mu}\Big(-\frac{GM}{r}\Big) \tag{2.65}$$

for a mass at a distance r from another (point) mass M. Comparison of this equation with the previous one yields

$$g_{00} \approx 1 - \frac{2GM}{r} . \tag{2.66}$$

In the general case we had $\vec{a} = -\vec{\nabla}\varphi$ and thus

$$\frac{1}{2}\vec{\nabla}g_{00} \approx \vec{\nabla}\varphi \tag{2.67}$$

so that g_{00} is just the Newtonian potential except for an additive constant and a factor two

$$g_{00} \approx 2\varphi + \text{const.} \tag{2.68}$$

• **Example—Around-the-World Atomic Clocks:** We consider a clock in an arbitrary gravitational field. It may be moving with arbitrary velocity, but not necessarily in free fall. If observed from a locally inertial frame ξ^α, then according to the equivalence principle the clock's rate is unaffected by the gravitational field and the spacetime interval between two ticks is given by

$$d\tau = \sqrt{\eta_{\alpha\beta}d\xi^\alpha d\xi^\beta} , \tag{2.69}$$

where

$$\eta_{\alpha\beta} = \begin{pmatrix} +1 & 0 & 0 & 0 \\ 0 & -1 & 0 & 0 \\ 0 & 0 & -1 & 0 \\ 0 & 0 & 0 & -1 \end{pmatrix} . \tag{2.70}$$

Here $d\tau$ is the time between two ticks if the clock is at <u>rest</u>. Now we transform to an arbitrary coordinate system. Hence

$$d\tau = \sqrt{\eta_{\alpha\beta}\frac{\partial\xi^\alpha}{\partial x^\mu}dx^\mu\frac{\partial\xi^\beta}{\partial x^\nu}dx^\nu} \,, \tag{2.71}$$

or

$$d\tau = \sqrt{g_{\mu\nu}dx^\mu dx^\nu} \tag{2.72}$$

(cf. (2.54)). For a clock whose velocity is dx^μ/dt the time between ticks is dt and we can write

$$\frac{dt}{d\tau} = \left(g_{\mu\nu}\frac{dx^\mu}{dt}\frac{dx^\nu}{dt}\right)^{-1/2}. \tag{2.73}$$

Inserting the above approximation, i.e.

$$g_{\mu\nu} \approx \begin{pmatrix} 1-\frac{2GM_\oplus}{r} & 0 & 0 & 0 \\ 0 & -1 & 0 & 0 \\ 0 & 0 & -1 & 0 \\ 0 & 0 & 0 & -1 \end{pmatrix} \tag{2.74}$$

yields

$$\frac{dt}{d\tau} \approx \left(1 - \frac{2GM_\oplus}{r} - v^2\right)^{-1/2}. \tag{2.75}$$

Now suppose we want to check this formula using accurate atomic clocks. We buy seats on a plane which takes several atomic clocks on a trip around the world (along the equator). Actually we can circle around the Earth in the same direction it rotates or opposite to its direction of rotation. Another set of atomic clocks remain stationary on the ground. They measure

$$\left.\frac{dt}{d\tau}\right|_g \approx \left(1 - \frac{2GM_\oplus}{R_\oplus} - (\omega R_\oplus)^2\right)^{-1/2}. \tag{2.76}$$

Note that $\omega R_\oplus$ is the velocity on the ground due to the rotation of the Earth whose radius is $R_\oplus$ and its angular velocity is ω. On the plane at altitude h and ground speed v_o the clocks measure

$$\left.\frac{dt}{d\tau}\right|_p \approx \left(1 - \frac{2GM_\oplus}{R_\oplus + h} - (\omega(R_\oplus + h) + v_o)^2\right)^{-1/2} \tag{2.77}$$

instead. To leading order the difference is

$$\left.\frac{dt}{d\tau}\right|_p - \left.\frac{dt}{d\tau}\right|_g \approx -\underbrace{\frac{GM_\oplus h}{R_\oplus^2}}_{1.09 \cdot 10^{-12}} + \underbrace{\frac{1}{2}v_o^2}_{0.35 \cdot 10^{-12}} + \underbrace{\omega R_\oplus v_o}_{1.29 \cdot 10^{-12}} . \tag{2.78}$$

Here we have used the values $h = 10^4$ m and $v_o = 900$ km/h (note: $GM_\oplus$ =0.00443 m and $R_\oplus = 6380$ km). Note that in this case the ground speed and the Earth's rotation are in the same direction. Otherwise v_o must be replaced by $-v_o$. In the former case (flight direction is east) a clock on the plane is behind of a ground clock by $\approx 0.55 \cdot 10^{-12}$ per tick and in the latter case (flight direction is west) it is ahead by $\approx 2.0 \cdot 10^{-12}$ per tick (as measured in the inertial frame of reference). An equatorial flight path at this ground speed takes about 44.6 hours. During this time the atomic clock on the plane heading east looses $0.55 \cdot 10^{-12} \cdot 44.6 \cdot 3600 \approx 88$ ns. The atomic clock on a plane heading west gains 317 ns. The gravitational effect by itself slows down the ground clock by 175 ns during the flight.

This *remarkable* experiment[a] was successfully carried out in 1971 by Haefele and Keating [5]. Haefele and Keating had noticed that atomic clock technology was sufficiently advanced. They booked seats for themselves as well as for their atomic clocks on commercial flights. Because there are no commercial flights around the world along the equator, they had to piece together different sections. Their numbers on the velocity effect therefore differ somewhat from ours.

---

[a]We consider this experiment '*remarkable*', because it made general relativity effects tangible.

• **Example—Testing Relativity Theory in the Laboratory via the Mössbauer Effect:** In 1958 Rudolf Mössbauer[a] discovered the recoilless nuclear resonance absorption of gamma radiation—the Mössbauer effect. With Fe^{57} this enables an energy resolution of $\Delta\nu/\nu \approx 10^{-12}$ (line width divided by the gamma energy). If we compare this number width the numerical values for the various quantities in (2.78) we notice that they are of the same order of magnitude. Of course, $h = 10$ km and $v = 900$ km/h, large numbers which enhance the effect, cannot be realised easily within a lab. Nevertheless, even the comparatively small values for h or v possible in the lab experiment can be sufficient to generate gamma frequency shifts detectable in a Mössbauer experiment. This was immediately recognised and quickly tried out.

One such experiment was performed by Pound and Rebka [6] (an earlier discussion of the possibility of this experiment appeared in [7]). The basic idea is to measure the absorption of photons parallel and anti-parallel to the direction of the Earth's gravitational field.

Imagine three observers. One observer watches the photons from a freely falling elevator and thus does not notice a gravitational field. More specifically, there is a mechanism which fires off the photons and simultaneously releases the elevator so that it can fall. From the perspective of the observer in the elevator there is no red- or blueshift, because there is no reason for it. The other two observers are stationary in the gravitational field. One is positioned at the bottom and the second at the top of the tower, which is used to carry out the experiment. According to (2.75) we have

$$\frac{dt_{\text{bottom}}}{d\tau} \approx \left(1 - \frac{2GM_{\oplus}}{R_{\oplus}}\right)^{-1/2}$$

and   (2.79)

$$\frac{dt_{\text{top}}}{d\tau} \approx \left(1 - \frac{2GM_{\oplus}}{R_{\oplus} + h}\right)^{-1/2}.$$

Here $v = 0$, because the two observers are at rest in the gravitational field. To leading order in h this yields

$$\frac{1}{dt_{\text{bottom}}} \approx \frac{1}{dt_{\text{top}}}(1 - gh).$$ (2.80)

Here $\frac{1}{dt_{\text{bottom}}}$ and $\frac{1}{dt_{\text{top}}}$ are the clock rates of the two observer clocks. Note that to at least first order in h we have

$$\frac{\omega_{\text{bottom}}}{dt_{\text{bottom}}} = \frac{\omega_{\text{top}}}{dt_{\text{top}}}.$$ (2.81)

and thus

$$\omega_{\text{bottom}} \approx \omega_{\text{top}}(1 + gh).$$ (2.82)

This is telling us that from the perspective of the observer in the elevator, the red- and blueshifting observed by the outside observers are due to their different clock rates determined by their respective heights in the gravitational field. For a generic potential difference $\delta\phi$ we thus have

$$\omega_{\text{bottom}} \approx \omega_{\text{top}} (1 + \delta\phi) . \tag{2.83}$$

The combined shift of both directions, i.e. photons emitted at the base of the tower and detected at a height $h = 22.6$ m above the base and vice versa, is $2gh \approx 4.9 \cdot 10^{-15}$. Even though this is at best 1% of the line width, the effect is measurable.

Another laboratory experiment using the Mössbauer effect in the context of relativity was carried out by Hay et al. [8]. These authors describe an experiment in which the absorber is mounted on a rotating disk at a distance R_a from the center. The plane of the disk is horizontal to the gravitational field, which consequently has no effect on the experiment. The emitter is at the center of the disk. We assume its radial position is R_e. Again we make use of (2.75), i.e.

$$\frac{dt_{R_e}}{d\tau} \approx \left(1 - \omega^2 R_e^2\right)^{-1/2}$$

$$\text{and} \tag{2.84}$$

$$\frac{dt_{R_a}}{d\tau} \approx \left(1 - \omega^2 R_a^2\right)^{-1/2} ,$$

where ω is the angular velocity. To leading order in the effect the clock rates differ by

$$\frac{1}{dt_{R_a}} - \frac{1}{dt_{R_e}} \approx -\frac{1}{2}\omega^2 \left(R_a^2 - R_e^2\right) . \tag{2.85}$$

What does this mean? Because $R_a > R_e$ we find that the clock rate of a clock travelling with the absorber is reduced (see problem 3). He sees the approaching photon blueshifted. Note that, due to the equivalence principle, the centripetal acceleration to this observer is equivalent to a gravitational field parallel to the direction of the approaching photon. The photon of course does not 'feel' a gravitational field.

Again, this is a difficult experiment. The large radius is about 6.6 cm, whereas the small radius is about 0.4 cm. The angular velocity, ω, is variable between 0 to 500 revolutions per second ($\omega = 500$ rev/sec leads to about 0.1% shift compared to the line width).

---

[a]Mössbauer, Rudolf, German physicist, *Munich 31.1.1929 , †Grünwald 14.9.2011; Nobel Prize in physics 1961.

2.4 Problems

1. Consider two people leisurely strolling by each other with a relative speed $v = 1.08$ km/h. According to one of the two, a nearby traffic light (distance 240 m) switches to green at exactly the same instant when they meet. If this person is walking in the direction of the traffic light and the other person exactly in the opposite direction, how much earlier or later does the traffic light switch in the reference frame of the other person? Now replace the switching traffic light with a supernova exploding in the Andromeda galaxy. How much earlier or later does it happen in the other persons reference frame? *Hint: Draw a spacetime diagram in which the first person is at rest.*

2. Imagine a 4 m long garage with a front and a back door that open and close automatically. The back door is initially open and closes instantaneously as soon as a car has completely passed it, while the front door is initially closed and opens instantaneously as soon as a car touches it. Now imagine a 4 m long car driving through the garage (back door to front door) at half the speed of light. Is it ever completely contained inside the garage, i.e. are both doors simultaneously shut at some point?

3. Consider a rigid disk of radius r spinning with constant angular velocity ω. There are two observers, A(lice) who is stationary and B(ob) who is at the edge of the disc, spinning with it.

   (a) What is the circumference of the disk according to Alice and according to Bob?

   (b) For Alice it takes a time T until she sees Bob pass infront of her again. How much time does it take for Bob?

4. In classical, Newtonian gravity the escape velocity from an object is the minimum velocity required to escape the gravitational binding to that object. Compute, for a point mass M, the escape velocity at a distance r. At what r does the escape velocity reach the speed of light c? Compute that radius r_s for a point mass equal to the mass of the Sun $M_\odot$. Ignore all relativistic effects in this calculation.

5. Every point in $1 + 1$ dimensional flat space (one time and one space coordinate) with the property $x > |t|$ can also be characterized by two coordinates ρ and α where ρ is the proper distance of the point from the origin and α characterizes how this point can be reached by a Lorentz transformation. Specifically, we have

$$t = \rho \sinh \alpha$$
$$x = \rho \cosh \alpha$$

   (a) Find the invariant line element $d\tau^2 = dt^2 - dx^2$ explicitly in the coordinates ρ and α. What is the metric $g_{\alpha\beta}$ in the new coordinates $y^0 = \alpha$ and $y^1 = \rho$?

   (b) What is the accelleration $a^\mu = d^2 x^\mu / d\tau^2$ and its norm $a_\mu a^\mu$ of a particle travelling along a line of constant ρ?

   (c) Find the Christoffel symbols $\Gamma^\mu_{\lambda\sigma}$ and write the equations of motion of a particle travelling in a straight line ($a^\mu = 0$!) in the new coordinates.

Chapter 3
Introduction to Multidimensional Calculus

Generalizing from flat to arbitrary spacetime, we develop the basic formalism of differential geometry, which is the mathematical foundation of the general theory of relativity. Vectors, tensors and the important concept of the covariant derivative are introduced. The latter allows us to compare vectors and tensors at different points. The generalization of symmetry transformations with Killing fields is also introduced.

3.1 Coordinate Transformation

Consider the scalar quantity $\phi = \phi(x)$ where $x = (x^1, x^2, \ldots, x^d)$. Note that the exponents are superscripts and d is the dimension of space. Thus

$$d\phi = \sum_{m=1}^{d} \frac{\partial \phi}{\partial x^m} dx^m . \tag{3.1}$$

From here on we use the summation convention, i.e.

$$d\phi = \frac{\partial \phi}{\partial x^m} dx^m . \tag{3.2}$$

Next we introduce the coordinate transformation

$$y^n = y^n(x) . \tag{3.3}$$

We assume that every point has a unique set of coordinates, i.e. there is a one-to-one correspondence of x and $y = (y^1, y^2, \ldots, y^d)$. How does ϕ change with y? The answer is

© Springer Nature Switzerland AG 2020
R. Hentschke and C. Hölbling, *A Short Course in General Relativity and Cosmology*, Undergraduate Lecture Notes in Physics,
https://doi.org/10.1007/978-3-030-46384-7_3

$$\frac{\partial \phi}{\partial y^n} = \frac{\partial x^m}{\partial y^n} \frac{\partial \phi}{\partial x^m} \; , \tag{3.4}$$

which can also be written without ϕ, i.e.

$$\boxed{\frac{\partial}{\partial y^n} = \frac{\partial x^m}{\partial y^n} \frac{\partial}{\partial x^m}} \; . \tag{3.5}$$

Finally, note that in (3.1) we may replace ϕ with a component of y, i.e.

$$\boxed{\mathrm{d}y^n = \frac{\partial y^n}{\partial x^m} \mathrm{d}x^m} \; . \tag{3.6}$$

The two formulas (3.5) and (3.6) is all we need to describe the transformation of tensors between reference frames.

3.2 Tensors

3.2.1 Transformation of Tensors

A vector whose components V_n transform according to (3.5), i.e.

$$V_n(y) = \frac{\partial x^m}{\partial y^n} V_m(x) \; , \tag{3.7}$$

is called a covariant vector. A vector whose components V^n transform according to (3.6), i.e.

$$V^n(y) = \frac{\partial y^n}{\partial x^m} V^m(x) \; , \tag{3.8}$$

is called a contravariant vector. The two types of vector components are distinguished by downstairs and upstairs index positions. Even though we use the same letter V, V_n is different from V^n! Note also that not every object made out of say three components is a co- or contravariant vector. Why? For instance, if one takes temperature as the first component, density as the second and humidity as the third, then this object transforms not like a vector in the above sense. Scalar quantities like temperature, density or humidity remain the same in different frames of reference.

The definitions (3.7) and (3.8) are generalisable to objects with several indices, e.g.

$$T_{np}(y) = \frac{\partial x^m}{\partial y^n} \frac{\partial x^q}{\partial y^p} T_{mq}(x) , \tag{3.9}$$

which is a covariant tensor of rank two. V_n and V^n are components of tensors of rank one. Analogously

$$T^{np}(y) = \frac{\partial y^n}{\partial x^m} \frac{\partial y^p}{\partial x^q} T^{mq}(x) , \tag{3.10}$$

is a contravariant tensor of rank 2. A mixed tensor is also possible, e.g.

$$T_p^n(y) = \frac{\partial y^n}{\partial x^m} \frac{\partial x^q}{\partial y^p} T_q^m(x) . \tag{3.11}$$

Notice the pattern! Indices maintain their positions (upstairs or downstairs). If they are summed over, then one index must be downstairs and the other one must be upstairs. We can add and subtract tensors, component by component, if they are of the same type.

• **Example—Transformation of the Affine Connection:** Starting from the special formula (2.52) for the affine connection, i.e.

$$\Gamma^\lambda_{\mu\nu} = \frac{\partial x^\lambda}{\partial \xi^\alpha} \frac{\partial^2 \xi^\alpha}{\partial x^\mu \partial x^\nu} ,$$

we want to work out its behavior under coordinate transformation. Passing from x^μ to a different system y^μ, we find

$$\begin{aligned}
\Gamma^\lambda_{\mu\nu}(y) &= \frac{\partial y^\lambda}{\partial \xi^\alpha} \frac{\partial^2 \xi^\alpha}{\partial y^\mu \partial y^\nu} \\
&= \frac{\partial y^\lambda}{\partial x^\rho} \frac{\partial x^\rho}{\partial \xi^\alpha} \frac{\partial}{\partial y^\mu} \left(\frac{\partial x^\sigma}{\partial y^\nu} \frac{\partial \xi^\alpha}{\partial x^\sigma} \right) \\
&= \frac{\partial y^\lambda}{\partial x^\rho} \frac{\partial x^\rho}{\partial \xi^\alpha} \left[\frac{\partial x^\sigma}{\partial y^\nu} \frac{\partial x^\tau}{\partial y^\mu} \frac{\partial^2 \xi^\alpha}{\partial x^\tau \partial x^\sigma} \right. \\
&\quad \left. + \frac{\partial^2 x^\sigma}{\partial y^\mu \partial y^\nu} \frac{\partial \xi^\alpha}{\partial x^\sigma} \right] .
\end{aligned}$$

This is

$$\Gamma^\lambda_{\mu\nu}(y) = \frac{\partial y^\lambda}{\partial x^\rho} \frac{\partial x^\tau}{\partial y^\mu} \frac{\partial x^\sigma}{\partial y^\nu} \Gamma^\rho_{\tau\sigma}(x) + \frac{\partial y^\lambda}{\partial x^\rho} \frac{\partial^2 x^\rho}{\partial y^\mu \partial y^\nu} . \tag{3.12}$$

Without the second term on the right hand side the transformation behavior would be that of a tensor. But this term makes $\Gamma^\lambda_{\mu\nu}$ a nontensor.

It will be useful to rewrite the second term in (3.12). Differentiating the identity

$$\frac{\partial y^\lambda}{\partial x^\rho}\frac{\partial x^\rho}{\partial y^\nu} = \delta^\lambda_\nu$$

with respect to y^μ yields

$$\frac{\partial y^\lambda}{\partial x^\rho}\frac{\partial^2 x^\rho}{\partial y^\mu \partial y^\nu} = -\frac{\partial x^\rho}{\partial y^\nu}\frac{\partial x^\sigma}{\partial y^\mu}\frac{\partial^2 y^\lambda}{\partial x^\rho \partial x^\sigma}$$

and therefore

$$\Gamma^\lambda_{\mu\nu}(y) = \frac{\partial y^\lambda}{\partial x^\rho}\frac{\partial x^\tau}{\partial y^\mu}\frac{\partial x^\sigma}{\partial y^\nu}\Gamma^\rho_{\tau\sigma}(x) - \frac{\partial x^\rho}{\partial y^\nu}\frac{\partial x^\sigma}{\partial y^\mu}\frac{\partial^2 y^\lambda}{\partial x^\rho \partial x^\sigma} . \tag{3.13}$$

3.2.2 The Metric Tensor

The distance, ds, between two infinitesimally close points in ordinary flat space is calculated via

$$ds^2 = \sum_{m=1}^{3} dx^m dx^m . \tag{3.14}$$

Using (3.6) we can compute ds in the y-frame, i.e.

$$ds^2 = \delta_{mn}\frac{\partial x^m}{\partial y^r}\frac{\partial x^n}{\partial y^s}dy^r dy^s . \tag{3.15}$$

Note that

$$\delta_{mn} = 1 \text{ if } m = n \text{ and } 0 \text{ otherwise} . \tag{3.16}$$

The quantity

$$g_{rs}(y) \equiv \delta_{mn}\frac{\partial x^m}{\partial y^r}\frac{\partial x^n}{\partial y^s} \tag{3.17}$$

is the metric tensor, the key object of Riemannian geometry. Note that the metric tensor is symmetric and that δ_{mn} is the metric tensor in cartesian coordinates. The general definition, not restricted to flat space, is

$$ds^2 = g_{mn}(x)dx^m dx^n \ . \tag{3.18}$$

We can see that g_{mn} indeed transforms as a covariant tensor via

$$\begin{aligned}
ds^2 &= g_{mn}(x)dx^m dx^n \\
&= \underbrace{g_{mn}(x)\frac{\partial x^m}{\partial y^r}\frac{\partial x^n}{\partial y^s}}_{=g_{rs}(y)} dy^r dy^s
\end{aligned} \tag{3.19}$$

Let's work out g_{rs} in a plane in polar coordinates, i.e. $x = r\cos\theta$ and $y = r\sin\theta$ (note: $x = x^1$, $y = x^2$ and $r = y^1$, $\theta = y^2$). Thus

$$\begin{aligned}
dx &= \frac{\partial x}{\partial r}dr + \frac{\partial x}{\partial \theta}d\theta = \cos\theta\,dr - r\sin\theta\,d\theta \\
dy &= \frac{\partial y}{\partial r}dr + \frac{\partial y}{\partial \theta}d\theta = \sin\theta\,dr + r\cos\theta\,d\theta
\end{aligned}$$

and

$$ds^2 = dr^2 + r^2 d\theta^2 \ . \tag{3.20}$$

The components of the metric tensor are

$$g_{rr} = 1 \quad g_{r\theta} = 0 \quad g_{\theta\theta} = r^2 \ . \tag{3.21}$$

Note that even though the curvilinear coordinates r and θ may convey the impression of curvature, we are still in flat space. As one can easily check, (3.17) is satisfied. Note also that $g_{r\theta} = 0$ means that lines of constant r and lines of constant ϕ are perpendicular.

Now let's work out g_{rs} on the surface of a unit sphere. In a three-dimensional cartesian coordinate system the surface of a unit sphere is swept out by the vector

$$\vec{z} = (\cos\phi\sin\theta, \sin\phi\sin\theta, \cos\theta) \ . \tag{3.22}$$

We construct a (locally) rectangular coordinate system tangential to the sphere's surface via the two tangent vectors

$$\begin{aligned}
d\vec{z}_\phi &= \frac{d\vec{z}}{d\phi}d\phi = (-\sin\phi\sin\theta, \cos\phi\sin\theta, 0)d\phi \\
d\vec{z}_\theta &= \frac{d\vec{z}}{d\theta}d\theta = (\cos\phi\cos\theta, \sin\phi\cos\theta, -\sin\theta)d\theta
\end{aligned}$$

in ϕ- and θ-direction, respectively. Thus

$$d\vec{z}_\phi{}^2 = \sin^2\theta d\phi^2 \quad \text{and} \quad d\vec{z}_\theta{}^2 = d\theta^2 \tag{3.23}$$

and

$$ds^2 = d\vec{z}_\phi{}^2 + d\vec{z}_\theta{}^2 = \sin^2\theta\, d\phi^2 + d\theta^2. \tag{3.24}$$

The sought after metric tensor components are

$$g_{\phi\phi} = \sin^2\theta \quad g_{\phi\theta} = 0 \quad g_{\theta\theta} = 1 . \tag{3.25}$$

Again, this looks as simple as (3.21). But this g_{rs} does not satisfy (3.17). To see this let's assume $x^1 = \sin\theta\,\phi$ and $x^2 = \theta$ according to (3.23)—and of course $y^1 = \phi$ and $y^2 = \theta$. What we get when we work out $g_{rs}(y)$ according to (3.17) however is quite different from (3.25). We can get (3.25) with a local version of x^m, by replacing $x^1 = \sin\theta\,\phi$ by $x^1 = \sin\theta_o\,\phi$, where θ_o is fixed. This will give us

$$g_{\phi\phi} = \sin^2\theta_o, \quad g_{\phi\theta} = 0 \quad g_{\theta\theta} = 1 . \tag{3.26}$$

We may look for other coordinate transformations $x = x(y)$, but we will not succeed in flattening the entire surface of the sphere. Only the local description of the distance between points on the sphere by cartesian coordinates is possible.

3.2.3 Raising and Lowering Indices

Because the metric tensor is symmetric, we can find its inverse $\mathbf{g}^{-1}$, i.e.

$$\mathbf{g}^{-1}\mathbf{g} = \mathbf{I} , \tag{3.27}$$

where $\mathbf{I}$ is the unit matrix. We write this in the following form:

$$\left(g^{-1}\right)^{mr} g_{rn} = \delta_n^m . \tag{3.28}$$

It is not obvious that the $\left(g^{-1}\right)^{mr}$ are components of a tensor and that it is a contravariant tensor. For the sake of brevity we are going to omit the -1 and simply use

$$g^{mr} := \left(g^{-1}\right)^{mr} . \tag{3.29}$$

Now we define

$$V^m g_{mn} := V_n \ .$$

(3.30)

At this point this really is a construction definition of how to make a V_n from a V^m. However, it is easy to show that if this definition holds in the y-frame it also is true in the x-frame:

$$V^m(y)g_{mn}(y) = \frac{\partial y^m}{\partial x^r} V^r(x)g_{vs}(x)\frac{\partial x^v}{\partial y^m}\frac{\partial x^s}{\partial y^n}$$
$$= \frac{\partial x^s}{\partial y^n} V^v(x)g_{vs}(x) \ .$$

and

$$V_n(y) = \frac{\partial x^s}{\partial y^n} V_s(x) \ .$$

This means that (3.30) is compatible with the concept of equations which hold true in all frames of reference. A special application of (3.30) is

$$ds^2 = g_{mn}dx^m dx^n = dx_n dx^n \ .$$

(3.31)

Applying g^{ns} to both sides of (3.30) and using (3.28) we find

$$V_n g^{ns} = V^s \ .$$

(3.32)

Generalisation of (3.30) and (3.32) to tensors of rank higher than one is straightforward, e.g.

$$T_{mn}g^{ns} = T_m^s \ .$$

(3.33)

• **Example—Equation of Motion and Conserved Quantities:** We start from (2.58), i.e.

$$\frac{du^\mu}{d\tau} + \Gamma^\mu_{\alpha\beta}u^\alpha u^\beta = 0 \ ,$$

(3.34)

where $u^\mu = dx^\mu/d\tau$ is the four-velocity. To this equation we apply $g_{\nu\mu}$, i.e.

$$g_{\nu\mu}\left(\frac{du^\mu}{d\tau} + \Gamma^\mu_{\alpha\beta}u^\alpha u^\beta\right) = 0 \ .$$

(3.35)

Note that

$$g_{\nu\mu}\left(\frac{du^{\mu}}{d\tau}\right) = \frac{d}{d\tau}\underbrace{\left(g_{\nu\mu}u^{\mu}\right)}_{=u_{\nu}} - u^{\mu}\underbrace{\frac{dx^{\sigma}}{d\tau}}_{=u^{\sigma}}\frac{\partial g_{\nu\mu}}{\partial x^{\sigma}} \tag{3.36}$$

and

$$g_{\nu\mu}\Gamma^{\mu}_{\alpha\beta} = \frac{1}{2}\delta^{\gamma}_{\nu}\left(\frac{\partial g_{\alpha\gamma}}{\partial x^{\beta}} + \frac{\partial g_{\gamma\beta}}{\partial x^{\alpha}} - \frac{\partial g_{\alpha\beta}}{\partial x^{\gamma}}\right)$$

$$= \frac{1}{2}\left(\frac{\partial g_{\alpha\nu}}{\partial x^{\beta}} + \frac{\partial g_{\nu\beta}}{\partial x^{\alpha}} - \frac{\partial g_{\alpha\beta}}{\partial x^{\nu}}\right). \tag{3.37}$$

Thus (3.35) becomes

$$\frac{du_{\nu}}{d\tau} = \frac{1}{2}\left(\frac{\partial g_{\alpha\beta}}{\partial x^{\nu}} + \frac{\partial g_{\alpha\nu}}{\partial x^{\beta}} - \frac{\partial g_{\nu\beta}}{\partial x^{\alpha}}\right)u^{\alpha}u^{\beta}. \tag{3.38}$$

Note that the last two terms in brackets are anti-symmetric in α and β, whereas $u^{\alpha}u^{\beta}$ is symmetric. Therefore (3.38) reduces to

$$\boxed{\frac{du_{\nu}}{d\tau} = \frac{1}{2}\frac{\partial g_{\alpha\beta}}{\partial x^{\nu}}u^{\alpha}u^{\beta}}. \tag{3.39}$$

A second form of this equation is obtained by expressing u^{α} in (3.38) via $u^{\alpha} = g^{\alpha\gamma}u_{\gamma}$. i.e.

$$\boxed{\frac{du_{\nu}}{d\tau} - \Gamma^{\gamma}_{\nu\beta}u_{\gamma}u^{\beta} = 0}. \tag{3.40}$$

Note that equation (3.39) tells us something important. If $g_{\alpha\beta}$ is independent of x^{ν} for some fixed index ν, then $du_{\nu}/d\tau = 0$, i.e. u_{ν} is a conserved quantity along the particle's trajectory.

3.3 Covariant Derivative

It is straightforward to compare the value of a scalar field at two different points. This property allows a unique definition of the derivative of a scalar field, its transformation between the x- and the y-coordinates (3.4) and ultimately the definition of a covariant vector (3.7). Comparing vectors at two different points however is not at all straightforward. To see why this is difficult, we have to talk about one property of vectors that we have swept under the rug until now: the space that they live in. Back in school we have learned the geometric picture that vectors are arrows possessing

Fig. 3.1 The vectors are the same in both sketches, In the curvilinear coordinaten system, however, their projections change from place to place

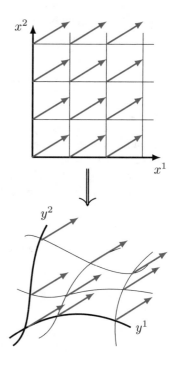

a certain length and pointing in some direction (the components of a contravariant vector are just the coordinates of that arrow). It might seem reasonable at first that each vector therefore points to another point in space—but this is not generally so. Since vectors give a direction at one specific point, they are tangent to the space in one specific point. If you picture your space to be a plain sheet of paper, the vectors drawn on it will point to other points in that space. But if you take the surface of a sphere instead, the tangent vectors at one point will live in a plane tangent to the sphere, the tangent space in that point. Comparing vectors at two different points is therefore difficult, because they do not even live in the same space! We have to specify how we map the tangent spaces at different points onto each other before we can compare vectors in them and define their derivatives.

Let's start comparing vectors at different points by just comparing their components. As illustrated in Fig. 3.1, the components of the vectors will depend on the choice of coordinates and thus we expect the derivative based on such a simple comparison to be coordinate dependent, i.e. not transform as a covariant tensor index. We first look at a covariant vector which obeys the transformation rule

$$V_\mu(y) = \frac{\partial x^\rho}{\partial y^\mu} V_\rho(x) \ . \tag{3.41}$$

Our simple prescription tells us to perform the derivative of the components with respect to y^ν yielding

$$\frac{\partial V_\mu(y)}{\partial y^\nu} = \frac{\partial x^\rho}{\partial y^\mu} \frac{\partial x^\sigma}{\partial y^\nu} \frac{\partial V_\rho(x)}{\partial x^\sigma} + \frac{\partial^2 x^\rho}{\partial y^\mu \partial y^\nu} V_\rho(x) . \tag{3.42}$$

As expected, the resulting expression does not transform as a covariant tensor. The second term on the right hand side destroys the tensor property of the derivative. It vanishes if x is linearly related to y (e.g. a rotation), but generically there will be an extra term. To get rid of this extra term we note that the transformation of Christoffel symbols $\Gamma^\lambda_{\mu\nu}(y)$ in (3.12) produces a similar term. Multiplying $\Gamma^\lambda_{\mu\nu}(y)$ onto (3.41) we obtain

$$\begin{aligned}
\Gamma^\lambda_{\mu\nu}(y) V_\lambda(y) &= \left[\frac{\partial y^\lambda}{\partial x^\kappa} \frac{\partial x^\tau}{\partial y^\mu} \frac{\partial x^\sigma}{\partial y^\nu} \Gamma^\kappa_{\tau\sigma}(x) \right. \\
&\quad \left. + \frac{\partial y^\lambda}{\partial x^\kappa} \frac{\partial^2 x^\kappa}{\partial y^\mu \partial y^\nu} \right] \frac{\partial x^\rho}{\partial y^\lambda} V_\rho(x) \\
&= \frac{\partial x^\rho}{\partial y^\mu} \frac{\partial x^\sigma}{\partial y^\nu} \Gamma^\kappa_{\rho\sigma}(x) V_\kappa(x) \\
&\quad + \frac{\partial^2 x^\kappa}{\partial y^\mu \partial y^\nu} V_\kappa(x) .
\end{aligned} \tag{3.43}$$

Subtracting (3.43) from (3.42) yields

$$\begin{aligned}
&\frac{\partial V_\mu(y)}{\partial y^\nu} - \Gamma^\lambda_{\mu\nu}(y) V_\lambda(y) \\
&= \frac{\partial x^\rho}{\partial y^\mu} \frac{\partial x^\sigma}{\partial y^\nu} \left(\frac{\partial V_\rho(x)}{\partial x^\sigma} - \Gamma^\kappa_{\rho\sigma}(x) V_\kappa(x) \right) .
\end{aligned} \tag{3.44}$$

This new quantity obviously has the sought after tensor property. We use it to define the covariant derivative of a covariant vector via

$$\boxed{\nabla_\nu V_\mu(x) := \frac{\partial V_\mu(x)}{\partial x^\nu} - \Gamma^\lambda_{\mu\nu}(x) V_\lambda(x)} . \tag{3.45}$$

And how do we find the covariant derivative of a contravariant vector V^μ? Instead of (3.41) we now have

$$V^\mu(y) = \frac{\partial y^\mu}{\partial x^\rho} V^\rho(x) . \tag{3.46}$$

This leads to

$$\frac{\partial V^\mu(y)}{\partial y^\nu} = \frac{\partial y^\mu}{\partial x^\rho} \frac{\partial x^\sigma}{\partial y^\nu} \frac{\partial V^\rho(x)}{\partial x^\sigma} + \frac{\partial x^\sigma}{\partial y^\nu} \frac{\partial^2 y^\mu}{\partial x^\sigma \partial x^\rho} V^\rho(x) \tag{3.47}$$

as compared to (3.42). Next we use (3.13) instead of (3.12), i.e.

$$\Gamma^\mu_{\nu\lambda}(y) V^\lambda(y)$$
$$= \frac{\partial y^\mu}{\partial x^\rho} \frac{\partial x^\tau}{\partial y^\lambda} \frac{\partial x^\sigma}{\partial y^\nu} \Gamma^\rho_{\tau\sigma}(x) \frac{\partial y^\lambda}{\partial x^\kappa} V^\kappa(x)$$
$$- \frac{\partial x^\rho}{\partial x^\nu} \frac{\partial x^\sigma}{\partial y^\lambda} \frac{\partial^2 y^\mu}{\partial x^\rho \partial x^\sigma} \frac{\partial y^\lambda}{\partial x^\kappa} V^\kappa(x)$$
$$= \frac{\partial y^\mu}{\partial x^\rho} \frac{\partial x^\sigma}{\partial y^\nu} \Gamma^\rho_{\kappa\sigma}(x) V^\kappa(x) - \frac{\partial x^\rho}{\partial x^\nu} \frac{\partial^2 y^\mu}{\partial x^\rho \partial x^\kappa} V^\kappa(x) \ . \tag{3.48}$$

Addition of (3.47) and (3.48) shows that

$$\boxed{\nabla_\nu V^\mu(x) := \frac{\partial V^\mu(x)}{\partial x^\nu} + \Gamma^\mu_{\nu\lambda}(x) V^\lambda(x)} \tag{3.49}$$

is the sought after covariant derivative of a contravariant vector.

How do we compute the covariant derivative of a tensor of rank 2? The answer is

$$\nabla_\lambda T_{\mu\nu} = \frac{\partial T_{\mu\nu}}{\partial y^\lambda} - \Gamma^\sigma_{\lambda\mu} T_{\sigma\nu} - \Gamma^\sigma_{\lambda\mu} T_{\mu\sigma} \ . \tag{3.50}$$

Aside from the ordinary derivative there are now two Γ-terms for each of the indices. Note that the components of V or T could be constant and still V or T have non-zero covariant derivatives. Note also that if we set $T_{\mu\nu} = V_\mu V_\nu$, then the above implies the product rule

$$\nabla_\lambda(V_\mu V_\nu) = V_\mu \nabla_\lambda V_\nu + V_\nu \nabla_\lambda V_\mu \ . \tag{3.51}$$

Of particular interest to us is the covariant derivative of the metric tensor. The equivalence principle allows us to introduce a coordinate system that is cartesian in the surrounding of an arbitrary point up to second derivative corrections.[1] Remember that in cartesian coordinates the metric tensor does not vary from point to point, which implies that $\nabla_\lambda g_{\mu\nu}(x) = 0$, when x are the local cartesian coordinates. If we now transform this equation to a y-coordinate system we still have

---

[1] In fact this choice is not unique. There are at least two frequently used local cartesian coordinate systems: Riemann normal coordinates and Fermi normal coordinates.

$$\nabla_\lambda g_{\mu\nu}(y) = 0 \ . \tag{3.52}$$

This is the nice thing about tensor equations—they remain valid in every coordinate system.

Before proceeding, let us introduce a compact and common notation for normal and covariant derivatives. We define

$$\boxed{\frac{\partial}{\partial x^\mu} = \partial_\mu \equiv {}_{,\mu}} \tag{3.53}$$

and

$$\boxed{\frac{\partial}{\partial x_\mu} = \partial^\mu \equiv {}^{,\mu}} \ . \tag{3.54}$$

An example for a double derivative is $\frac{\partial^2}{\partial x^\nu \partial x^\mu} \equiv {}_{,\mu\nu}$. Similarly, for covariant derivatives we use the semicolon instead of the comma, i.e.

$$\nabla_\mu \equiv {}_{;\mu} \quad \text{and} \quad \nabla x^\mu \equiv {}^{;\mu} \ . \tag{3.55}$$

In the following we may alternate between notations.[2]

3.3.1   Covariant Derivative Along a Curve

We need to discuss one more type of covariant derivative, i.e. the covariant derivative along a curve parametrised by τ. Consider a convariant vector $V^\mu(y(\tau))$ with the transformation rule

$$V^\mu(y(\tau)) = \frac{\partial y^\mu}{\partial x^\nu} V^\nu(x(\tau)) \ . \tag{3.56}$$

Taking the derivative with respect to τ we find

---

[2]Note that in flat spacetime we have the relations $\frac{\partial}{\partial x^\mu} = (\partial_t, \partial_x, \partial_y, \partial_z) = \partial_\mu \equiv {}_{,\mu}$ and $\frac{\partial}{\partial x_\mu} = (\partial_t, -\partial_x, -\partial_y, -\partial_z) = \partial^\mu \equiv {}^{,\mu}$.

$$\frac{dV^\mu(y(\tau))}{d\tau} = \frac{\partial y^\mu}{\partial x^\nu} \frac{dV^\nu(x(\tau))}{d\tau}$$
$$+ \left(\frac{dx^\lambda}{d\tau} \frac{\partial^2 y^\mu}{\partial x^\lambda \partial x^\nu} \right) V^\nu(x(\tau)) .$$

(3.57)

It is not unexpected that we find a result similar to (3.42). We can proceed analogously and look at the transformation of $\Gamma^\mu_{\nu\lambda}(y(\tau))V^\nu(y(\tau))\frac{dy^\lambda}{d\tau}$. Using (3.12) we obtain

$$\Gamma^\mu_{\nu\lambda}(y)V^\nu(y)\frac{dy^\lambda}{d\tau}$$
$$= \frac{\partial y^\mu}{\partial x^\rho}\Gamma^\rho_{\nu\lambda}(x)V^\nu(x)\frac{dx^\lambda}{d\tau}$$
$$- \left(\frac{dx^\lambda}{d\tau} \frac{\partial^2 y^\mu}{\partial x^\lambda \partial x^\nu} \right) V^\nu(x) .$$

(3.58)

Adding (3.57) and (3.58) yields

$$\frac{dV^\mu(y)}{d\tau} + \Gamma^\mu_{\nu\lambda}(y)V^\nu(y)\frac{dy^\lambda}{d\tau}$$
$$= \frac{\partial y^\mu}{\partial x^\nu}\left(\frac{dV^\nu(x)}{d\tau} + \Gamma^\nu_{\sigma\lambda}(x)V^\sigma(x)\frac{dx^\lambda}{d\tau} \right) .$$

(3.59)

This is the desired tensor behavior under transformation and we define the covariant derivative of a contravariant vector along a curve via

$$\boxed{\nabla_\tau V^\nu(x) := \frac{dV^\nu(x)}{d\tau} + \Gamma^\nu_{\sigma\lambda}(x)V^\sigma(x)\frac{dx^\lambda}{d\tau} .}$$

(3.60)

Notice the close analogy between this equation and (3.49). The same similarity applies between (3.45) and the covariant derivative of a covariant vector along a curve.

Defining the velocity $u^\mu = dx^\mu/d\tau$ along the curve, we see that

$$\nabla_\tau V^\nu(x) = u^\nu \nabla_\nu V^\nu(x) .$$

(3.61)

Finally, applying this identity to the velocity itself allows us to recast the equation of motion (2.58) into the simple form

$$\nabla_\tau u^\mu = u^\nu \nabla_\nu u^\mu = 0 .$$

(3.62)

Fig. 3.2 The surface of a cone is rotationally symmetric around its axis. We can find this symmetry without leaving the cone by finding a field of Killing vectors (depicted here in blue). If an infinitesimal transformation generated by the vectors leaves the metric unchanged, we have found a symmetry

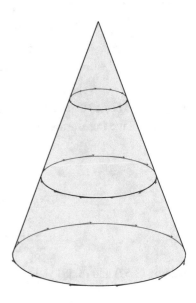

3.4   Symmetries of Curved Spacetime

The final concept we need to generalize is that of a symmetry. Let us look at the cone depicted in Fig. 3.2. Obviously, the cone is symmetric with respect to a rotation around its axis. But how do we formulate this if we live on the surface of that cone? The answer is, we introduce a vector field that does the rotation. At every point x on the cone's mantle we have an infinitesimal vector $\xi^\mu d\lambda$ that tells us into which neighbouring point y the point x is transformed. If after the transformation the metric is the same, the vector field describes a symmetry transformation and is called a Killing vector.[3]

But how do we identify Killing vectors? Let us start by introducing an arbitrary path $x^\mu(\lambda)$ parameterised by λ. If we start at a point x^μ and continue an infinitesimal distance $d\lambda$ along this path, we end up in a point

$$y^\mu = x^\mu(\lambda + d\lambda) = x^\mu(\lambda) + \xi^\mu d\lambda \qquad \xi^\mu = \frac{dx^\mu}{d\lambda} . \tag{3.63}$$

We can view this little shift in two ways: Either we can see it as just a change of coordinates. In that case, we can write for any vector

---

[3] Killing, Wilhelm Karl Joseph, German mathematician, *Burbach 10.5.1847, †Münster 11.2.1923.

$$A^\mu(y) = A^\mu(x + \xi d\lambda)$$
$$= A^\mu(x) + \xi^\alpha \partial_\alpha A^\mu(x) d\lambda$$

(3.64)

$$A_\mu(y) = A_\mu(x + \xi d\lambda)$$
$$= A_\mu(x) + \xi^\alpha \partial_\alpha A_\mu(x) d\lambda$$

and for the metric tensor

$$g_{\mu\nu}(y) = g_{\mu\nu}(x + \xi d\lambda)$$
$$= g_{\mu\nu}(x) + \xi^\alpha \partial_\alpha g_{\mu\nu}(x) d\lambda .$$

(3.65)

Alternatively, we might think of the little shift as a coordinate transformation from x^μ to y^μ. Viewed in that way, we have

$$\frac{\partial y^\alpha}{\partial x^\beta} = \delta^\alpha_\beta + \partial_\beta \xi^\alpha d\lambda$$

(3.66)

and thus, according to (3.7) and (3.8) vectors transform as

$$A'^\mu(y) = \frac{\partial y^\mu}{\partial x^\alpha} A^\alpha(x)$$
$$= A^\mu(x) + A^\alpha(x) \partial_\alpha \xi^\mu d\lambda$$

(3.67)

$$A'_\mu(y) = \frac{\partial x^\alpha}{\partial y^\mu} A_\alpha(x)$$
$$= A_\mu(x) - A_\alpha(x) \partial_\mu \xi^\alpha d\lambda$$

and the metric tensor

$$g'_{\mu\nu}(y) = \frac{\partial x^\alpha}{\partial y^\mu} \frac{\partial x^\beta}{\partial y^\nu} g_{\alpha\beta}(x)$$
$$= g_{\mu\nu}(x) - g_{\alpha\nu}(x) \partial_\mu \xi^\alpha d\lambda - g_{\mu\beta}(x) \partial_\nu \xi^\beta d\lambda .$$

(3.68)

The difference between the two interpretations allows us to form the Lie derivative[4]

$$\pounds_\xi A^\mu = \frac{A^\mu(y) - A'^\mu(y)}{d\lambda} = \xi^\alpha \partial_\alpha A^\mu - A^\alpha \partial_\alpha \xi^\mu$$

$$\pounds_\xi A_\mu = \frac{A_\mu(y) - A'_\mu(y)}{d\lambda} = \xi^\alpha \partial_\alpha A_\mu + A_\alpha \partial_\mu \xi^\alpha$$

(3.69)

$$\pounds_\xi g_{\mu\nu} = \frac{g_{\mu\nu}(y) - g'_{\mu\nu}(y)}{d\lambda} = \xi^\alpha \partial_\alpha g_{\mu\nu} + g_{\alpha\nu} \partial_\mu \xi^\alpha + g_{\mu\beta} \partial_\nu \xi^\beta .$$

---

[4]Lie, Sophus, Norwegian mathematician, *Nordfjordeit 17.12.1842, †Kristiania 18.2.1899.

A vanishing Lie derivative of a tensor tells us that this tensor sees no difference between the little shift of spacetime we made and a simple coordinate transformation. If the metric tensor does not see this difference, the little shift is a symmetry and the corresponding vector field ξ^μ that generates this symmetry transformation is a Killing vector. So the vector field ξ^μ is a Killing vector if

$$\pounds_\xi g_{\mu\nu} = 0 . \tag{3.70}$$

This is equivalent to (see Problem 4)

$$\nabla_\mu \xi_\nu + \nabla_\nu \xi_\mu = 0 \tag{3.71}$$

which is referred to as the Killing equation.

The utility of a Killing vector becomes apparent when we look at a geodesic $x^\mu(\tau)$ with the velocity $u^\mu = dx^\mu/d\tau$. We form the product of Killing vector and velocity and see how it changes along the geodesic, i.e.

$$\nabla_\tau (\xi_\mu u^\mu) = u^\mu \nabla_\tau \xi_\mu + \xi_\mu \nabla_\tau u^\mu . \tag{3.72}$$

The second term on the right hand side vanishes because of the equation of motion (3.62). The first term may be written as

$$\begin{aligned} u^\mu \nabla_\tau \xi_\mu &= u^\mu u^\nu \nabla_\nu \xi_\mu \\ &= -u^\mu u^\nu \nabla_\mu \xi_\nu \end{aligned} \tag{3.73}$$

where we have used the Killing equation (3.71) to obtain the last line. This is both symmetric and antisymmetric upon exchanging of μ and ν and thus also vanishes. We thus obtain

$$\nabla_\tau (\xi_\mu u^\mu) = 0 , \tag{3.74}$$

i.e. $\xi_\mu u^\mu$ is a conserved quantity along the geodesic. This generalizes the example we have given in Sect. 3.2.3.

3.5 Problems

1. Here we want to investigate a bit why Newtonian gravity cannot be simply included into special relativity.

   (a) Rewrite the Newtonian equation of motion in covariant form

   $$F^\mu = m \frac{d}{d\tau} \frac{dx^\mu}{d\tau}$$

and show that $d\tau^2 = g_{\mu\nu}dx^\mu dx^\nu$ then implies

$$F_\mu \frac{dx^\mu}{d\tau} = 0 .$$

(b) Write down a covariant version of the equation

$$\vec{F} = -m\vec{\nabla}\varphi$$

Using the relation derived in (a), show that the potential is constant along any particle trajectory and hence there is no gravitational force.

2. The points on a cone mantle in 3-dimensional cartesian coordinates are given as

$$\begin{pmatrix} x \\ y \\ z \end{pmatrix} = r \begin{pmatrix} \sin\theta\cos\varphi \\ \sin\theta\sin\varphi \\ \cos\theta \end{pmatrix}$$

with a constant cone angle 2θ. The two coordinates are $r \in [0, s]$ the distance from the tip of the cone and $\varphi \in [0, 2\pi]$ the angular coordinate on the conic surface.

(a) Compute the metric g_{ab} and the Christoffel symbols Γ^a_{bc} in the coordinates r and φ. *Hint: start from the invariant line element* $ds^2 = dx^2 + dy^2 + dz^2 = g_{ab}dy^a dy^b$ *with* $y^1 = r$ *and* $y^2 = \varphi$. *Also, if you happen to find a more convenient second coordinate by rescaling* φ *by a constant factor, you may use it for the rest of the exercise.*
(b) Find a coordinate transformation from r, φ to two-dimensional cartesian coordinates X, Y. If you want, you may picture this as the unrolling of the mantle into a plane and putting a cartesian coordinate grid onto the plane. *Hint: two dimensional cartesian coordinates fulfill* $ds^2 = dX^2 + dY^2$.
(c) A unit contravariant vector that points away from the tip of the cone is given by

$$y^a = \begin{pmatrix} r \\ \varphi \end{pmatrix} = \begin{pmatrix} 1 \\ 0 \end{pmatrix}$$

in (r, φ) coordinates. Compute the corresponding covariant vector y_a, the covariant derivative $y_{a;b}$ and the difference $y_{a;b} - y_{b;a}$. Write down the same vector in two-dimensional cartesian coordinates

$$X^a = \begin{pmatrix} X \\ Y \end{pmatrix}$$

and find X_a, $X_{a;b}$ and $X_{a;b} - X_{b;a}$.
(d) Show that the vector

$$\xi^a = \begin{pmatrix} r \\ \varphi \end{pmatrix} = \begin{pmatrix} 0 \\ 1 \end{pmatrix}$$

is a Killing vector.

3. The points on a spherical surface in 3-dimensional cartesian coordinates are given as

$$\begin{pmatrix} x \\ y \\ z \end{pmatrix} = r \begin{pmatrix} \sin\theta\cos\varphi \\ \sin\theta\sin\varphi \\ \cos\theta \end{pmatrix}$$

with a constant radius r of the sphere and the usual spherical coordinates $\theta \in [0, \pi]$ and $\varphi \in [0, 2\pi)$.

(a) Compute the metric g_{ab} and the Christoffel symbols Γ^a_{bc} in the coordinates θ and φ. *Hint: start form the invariant line element* $ds^2 = dx^2 + dy^2 + dz^2 = g_{ab}dy^a dy^b$ *with* $y^1 = \theta$ *and* $y^2 = \varphi$.

(b) We now look at a contravariant vector in (θ, φ) coordinates

$$y^a = \begin{pmatrix} \theta \\ \varphi \end{pmatrix} = \begin{pmatrix} 0 \\ r^{-1}\sin^{-1}\theta \end{pmatrix}.$$

Compute the corresponding covariant vector y_a and show that y^a has unit norm. Compute the covariant derivative $y_{a;b}$ and the difference $y_{a;b} - y_{b;a}$.

4. Prove that the Lie derivative of a covariant/contravariant vector is itself a covariant/contravariant vector. Use this result to show that the Killing equation $\nabla_\mu \xi_\nu + \nabla_\nu \xi_\mu = 0$ follows from $\pounds_\xi g_{\mu\nu} = 0$.

5. Show that $\sqrt{-g}\, d^4x$, where $g = \det g_{\mu\nu}$, is an invariant volume element.

Chapter 4
Field Equations of General Relativity

The core of general relativity are the Einstein field equations, which relate curvature to energy and momentum. This chapter introduces curvature and the energy-momentum tensor to derive tensor equations between them that fulfill the correct Newtonian limit—the Einstein field equation. Finally, the small curvature limit in which the Einstein field equations are linear is worked out. This establishes the contact with the other classical field theory, Maxwell electrodynamics.

4.1 Curvature

4.1.1 Parallel Transport

At this point we do have all the tools necessary to begin the construction of Einstein's field equation. But before doing this we must understand curvature.

Figure 4.1 depicts a cylindrical tube with a red circle around its circumference. We take a pair of scissors and cut along the dotted line. The cylindrical tube then becomes a flat rectangular sheet. At point α we draw an arrow parallel to the edge of the sheet, followed by additional arrows of the same length. Each new arrow is strictly parallel to its predecessor. The last arrow is drawn at point ω. Subsequently we roll the sheet up into its original shape—the cylindrical tube. If we compare the arrows at positions α and ω along the red line, we note that they are parallel. No surprise.

The bottom part of the same figure shows a cone. We want to repeat the above procedure for the cone. There is a red line on the surface of the cone with the two adjacent points α and ω. We take a pair of scissors and cut from the base to the tip along the dashed-dotted line. The result is a flat circular sheet with a wedge-like piece missing. The angle of the missing wedge is θ. Now we draw arrows as before. We begin with an arrow at α, which is nearly perpendicular to the edge of the missing

© Springer Nature Switzerland AG 2020
R. Hentschke and C. Hölbling, *A Short Course in General Relativity and Cosmology*, Undergraduate Lecture Notes in Physics,
https://doi.org/10.1007/978-3-030-46384-7_4

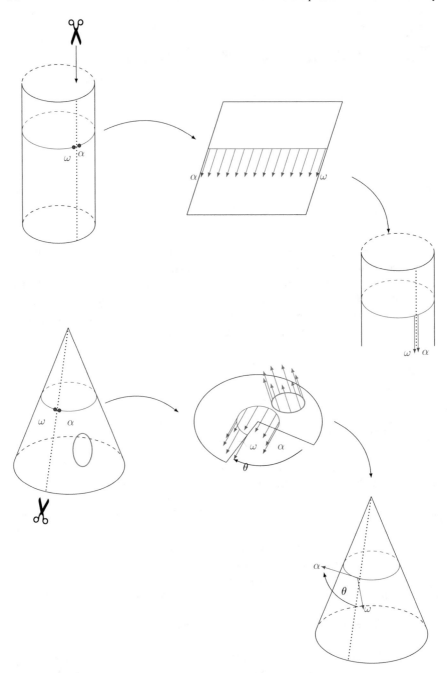

Fig. 4.1 Parallel transport of arrows on the surfaces of cylinders and cones

Fig. 4.2 Curvature of a spherical surface

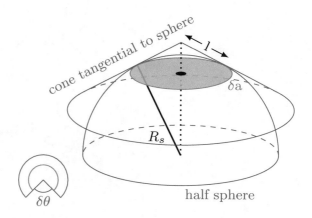

piece. More arrows follow, always parallel to the previous arrow. Finally we reach point ω on the other side of the missing wedge. Here we note that this arrow is not really perpendicular to the edge. If we now 'repair' the cone, we find that the α- and the ω-arrow are not parallel at all. The angle between them is θ.

We note that such an 'angular deficit' occurs only when a path surrounds the tip. The 'parallel transport' of arrows along other paths, like the green one, yields no angular deficit. This is because the cone's surface, as the cylinder's surface, is flat—except at the tip. The tip has curvature. Because the tip is sharp, we may wonder whether the method of parallel transport for measuring intrinsic curvature can be applied to smoother geometries.

Figure 4.2 shows a hemisphere with a cone-shaped hat. The points were cone and sphere are in contact form a small (green) circle enclosing the area δa. From the point of view of the cone the circle is just a path around its tip giving rise to an angular deficit $\delta\theta$. Note that if δa is small, this means that the cone is quite flat already. Cutting the cone yields a narrow missing wedge, i.e. $\delta\theta$ is also small. But there is more here. Along the green circle the cone's surface is tangential to the sphere's. The parallel transport of an arrow takes place not only in the cone's surface but also in the tangential space of the sphere. Therefore the angular deficit is not only that of the cone but also that of the sphere for the particular path.

It is an easy exercise (see Problem 1) to show that

$$\delta\theta = \frac{1}{R_s^2}\delta a , \qquad (4.1)$$

where R_s is the radius of the sphere. More generally

$$\boxed{\delta\theta = \mathcal{R}\,\delta a} , \qquad (4.2)$$

$\mathcal{R}$ is the (intrinsic) curvature. Note that curvature can be positive or negative. The rule is as follows. We surround the path counter-clockwise from α to ω. The sign

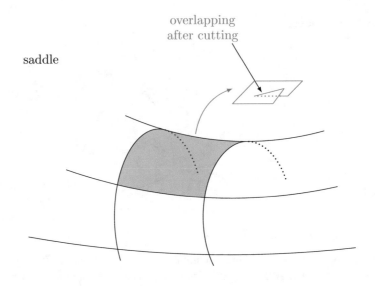

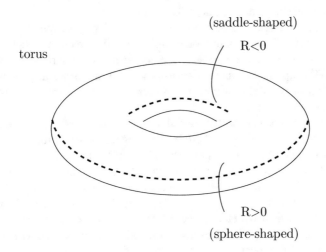

Fig. 4.3 Saddle and torus

of the angle $\delta\theta$ is positive, and so is the curvature, if we must rotate the α-arrow counter-clockwise towards the ω-arrow. Therefore the sphere's surface has a positive curvature. An example for a surface with negative curvature is the saddle shown in Fig. 4.3. Another example in this figure is the torus, whose surface has both positive and negative curvature.

One important *Remark* is in order. The arrows which get transported are never protruding from the surface. They 'live' entirely in the tangent space of the surface and do not have a component orthogonal to it. In this sense the curvature we have

Fig. 4.4 Closed path in the x^ν-x^σ-plane

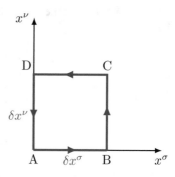

defined is intrinsic: it never refers to the space we have embedded our surface in. We now must devise a mathematical procedure for the measurement of curvature in an arbitrary space which generalises these examples.

4.1.2 The Riemann Tensor

The general procedure works as follows. Pick two axes x^ν and x^σ and parallel transport a vector V^μ along an infinitesimal closed path spanned by the two axes (cf. Fig. 4.4). You can now compare the parallel transported vector $\hat{V}^\mu$ to the original one and if there is curvature, they will not be the same. The difference $\hat{V}^\mu - V^\mu$ will thus be proportional to the original vector V^μ and the infinitesimal paths along both axes δx^ν and δx^σ with the curvature as the proportionality constant

$$\hat{V}^\mu - V^\mu = \oint_{\text{path}} dV^\mu = \mathcal{R}^\mu_{\tau\nu\sigma} \delta x^\nu \delta x^\sigma V^\tau . \tag{4.3}$$

Since there are four separate directions involved—the two axes, the original vector and the difference of the parallel transported vector to the original one—the curvature R has to have four indices. And because all other quantities are contravariant vectors, we thus have a rank four curvature tensor, also called the Riemann tensor $\mathcal{R}^\mu_{\tau\sigma\nu}$, encoding the complete information about curvature.

Let us now compute the integral. We will do so by following the path designated in Fig. 4.4, i.e. we will parallel transport the vector V^μ along an infinitesimal rectangle. First we note that generically when we parallel transport a vector V^μ along an axis δx^σ, its covariant derivative vanishes, i.e. $\nabla_\sigma V^\mu = \partial_\sigma V^\mu - \Gamma^\mu_{\sigma\tau} V^\tau = 0$. This implies a change in V^μ

$$dV^\mu(x) = -\Gamma^\mu_{\sigma\tau}(x) V^\tau(x) dx^\sigma . \tag{4.4}$$

Focussing first on the sum of the pieces AB and CD along the path in Fig. 4.4 we have

$$\delta V^\mu(x^\nu) + \delta V^\mu(x^\nu + \delta x^\nu) =$$
$$- \left(\Gamma^\mu_{\sigma\tau}(x^\nu) V^\tau(x^\nu) \right.$$
$$\left. - \Gamma^\mu_{\sigma\tau}(x^\nu + \delta x^\nu) V^\tau(x^\nu + \delta x^\nu) \right) \delta x^\sigma$$
$$= \frac{\partial}{\partial x^\nu} \left(\Gamma^\mu_{\sigma\tau}(x^\nu) V^\tau(x^\nu) \right) \delta x^\nu \delta x^\sigma \ .$$

The arguments x^ν and $x^\nu + \delta x^\nu$ in V^μ and $\Gamma^\mu_{\sigma\tau}$ are merely meant as orientation as to where along the path we are. Analogously the sum of the pieces BC and AD is

$$\delta V^\mu(x^\sigma + \delta x^\sigma) + \delta V^\mu(x^\sigma) =$$
$$= -\frac{\partial}{\partial x^\sigma} \left(\Gamma^\mu_{\nu\tau}(x^\sigma) V^\tau(x^\sigma) \right) \delta x^\nu \delta x^\sigma \ .$$

Hence

$$\oint_{\text{path}} dV^\mu = \left(\frac{\partial(\Gamma^\mu_{\sigma\tau} V^\tau)}{\partial x^\nu} - \frac{\partial(\Gamma^\mu_{\nu\tau} V^\tau)}{\partial x^\sigma} \right) \delta x^\nu \delta x^\sigma$$
$$\overset{(4.4)}{=} \left(\frac{\partial \Gamma^\mu_{\sigma\tau}}{\partial x^\nu} - \Gamma^\mu_{\sigma\rho} \Gamma^\rho_{\nu\tau} - \frac{\partial \Gamma^\mu_{\nu\tau}}{\partial x^\sigma} + \Gamma^\mu_{\nu\rho} \Gamma^\rho_{\sigma\tau} \right)$$
$$\delta x^\nu \delta x^\sigma V^\tau \ .$$

Comparison with (4.3) yields the Riemann tensor

$$\boxed{\mathcal{R}^\mu_{\ \tau\nu\sigma} = \frac{\partial \Gamma^\mu_{\sigma\tau}}{\partial x^\nu} - \frac{\partial \Gamma^\mu_{\nu\tau}}{\partial x^\sigma} + \Gamma^\mu_{\nu\rho} \Gamma^\rho_{\sigma\tau} - \Gamma^\mu_{\sigma\rho} \Gamma^\rho_{\nu\tau}} \ . \qquad (4.5)$$

Note that we already have (2.59), i.e. the Christoffel symbols in terms of the metric tensor. This means that now we express the Riemann tensor in terms of the the metric tensor.

The Riemann tensor

$$\mathcal{R}_{\lambda\tau\nu\sigma} = g_{\lambda\mu} \mathcal{R}^\mu_{\ \tau\nu\sigma} \qquad (4.6)$$

has some interesting properties which follow directly from the properties of the Christoffel symbols. In order to derive them we again invoke locally flat coordinates, where at any point we can have a Lorentz metric with all first derivatives (and thus Christoffel symbols) vanishing. In the expression (4.5) for the Riemann tensor only the derivative terms survive and thus we find

$$\boxed{\mathcal{R}_{\lambda\tau\nu\sigma} = \frac{1}{2}(g_{\lambda\sigma,\tau\nu} + g_{\tau\nu,\lambda\sigma} - g_{\sigma\tau,\lambda\nu} - g_{\lambda\nu,\tau\sigma})} . \tag{4.7}$$

In this local form, the symmetries of the Riemann tensor are evident and since it is a tensor, they remain true in an arbitrary coordinate system. Specifically we have:

(A) Symmetry:

$$\mathcal{R}_{\lambda\mu\nu\kappa} = \mathcal{R}_{\nu\kappa\lambda\mu} \tag{4.8}$$

(B) Antisymmetry:

$$\mathcal{R}_{\lambda\mu\nu\kappa} = -\mathcal{R}_{\mu\lambda\nu\kappa} = -\mathcal{R}_{\lambda\mu\kappa\nu} = \mathcal{R}_{\mu\lambda\kappa\nu} \tag{4.9}$$

(C) Cyclicity:

$$\mathcal{R}_{\lambda\mu\nu\kappa} + \mathcal{R}_{\lambda\kappa\mu\nu} + \mathcal{R}_{\lambda\nu\kappa\mu} = 0 \tag{4.10}$$

Contraction of the first and third index yields the Ricci tensor

$$\boxed{\mathcal{R}_{\mu\kappa} = g^{\lambda\nu}\mathcal{R}_{\lambda\mu\nu\kappa}} , \tag{4.11}$$

which, because of A, is symmetric. Property (B) tells us that $\mathcal{R}_{\mu\kappa}$ is essentially the only second-rank tensor that can be formed from the Riemann tensor.[1] The symmetry property (B) also leads to the conclusion that the only contraction of the Riemann tensor to a scalar, the Ricci scalar, is

$$\boxed{\mathcal{R} = g^{\lambda\nu}g^{\mu\kappa}\mathcal{R}_{\lambda\mu\nu\kappa}} . \tag{4.12}$$

In addition the curvature tensor satisfies the important Bianchi identity (see Problem 2), i.e.

$$\nabla_\eta \mathcal{R}_{\lambda\mu\nu\kappa} + \nabla_\kappa \mathcal{R}_{\lambda\mu\eta\nu} + \nabla_\nu \mathcal{R}_{\lambda\mu\kappa\eta} = 0 . \tag{4.13}$$

Remembering that the covariant derivative obeys the product rule (cf. (3.51)) and that $\nabla_\eta g^{\lambda\nu} = 0$, we find on contraction of λ with ν that

$$\nabla_\eta \mathcal{R}_{\mu\kappa} - \nabla_\kappa \mathcal{R}_{\mu\eta} + \nabla_\nu \mathcal{R}^\nu_{\mu\kappa\eta} = 0 . \tag{4.14}$$

Contracting again yields

---

[1] Multiplying (4.9) by $g^{\lambda\nu}$, $g^{\lambda\mu}$, and $g^{\nu\kappa}$ yields $\mathcal{R}_{\mu\kappa} = -g^{\lambda\nu}\mathcal{R}_{\mu\lambda\nu\kappa} = -g^{\lambda\nu}\mathcal{R}_{\lambda\mu\kappa\nu} = g^{\lambda\nu}\mathcal{R}_{\mu\lambda\kappa\nu}$ and $g^{\lambda\mu}\mathcal{R}_{\lambda\mu\nu\kappa} = g^{\nu\kappa}\mathcal{R}_{\lambda\mu\nu\kappa} = 0$.

$$\nabla_\eta \mathcal{R} - \nabla_\mu \mathcal{R}^\mu_\eta - \nabla_\nu \mathcal{R}^\nu_\eta = 0 \tag{4.15}$$

or

$$\nabla_\mu \left(\mathcal{R}^\mu_\eta - \frac{1}{2} \delta^\mu_\eta \mathcal{R} \right) = 0 \tag{4.16}$$

or equivalently

$$\nabla_\mu \left(\mathcal{R}^{\mu\nu} - \frac{1}{2} g^{\mu\nu} \mathcal{R} \right) = 0 \,. \tag{4.17}$$

4.2   Energy-Momentum Tensor

The other half of gravity is mass. The first one is curvature. Mass affects geometry, i.e. mass curves geometry. Here we want to define some notions important in this context.

Let's return to space and talk briefly about electric current. In electrodynamics we had defined the four vector

$$(\rho, j^x, j^y, j^z) \rightarrow j^\mu \,. \tag{4.18}$$

Here ρ is the density of charge q and the j^i ($i = 1, 2, 3$) are the current densities in the respective directions, i.e. j^x means number of charges q per area and per time in x-direction. Charge is a conserved quantity. It therefore fulfils the continuity equation

$$\frac{\partial \rho}{\partial t} + \vec{\nabla} \cdot \vec{j} = 0 \,. \tag{4.19}$$

Expressed in the four-notation this becomes

$$\frac{\partial j^0}{\partial x^0} + \frac{\partial j^1}{\partial x^1} + \frac{\partial j^2}{\partial x^2} + \frac{\partial j^3}{\partial x^3} = 0 \tag{4.20}$$

or

$$\frac{\partial j^\mu}{\partial x^\mu} = 0 \,. \tag{4.21}$$

Now let's go to energy and momentum. Thus q becomes E, p^1, p^2, and p^3. Energy and momentum are conserved, but they are not invariants. They look different in different frames of reference. For example, a mass may be at rest in one frame of reference and therefore has no kinetic energy. From the perspective of a moving frame of reference this is different.

In the following

$$(E, \vec{p}) \rightarrow (\underbrace{p^0, p^{1,2,3}}_{\text{quantities}}) \tag{4.22}$$

and all components are conserved. Again there is a flow of energy and momentum according to the following scheme:

$$\underbrace{\mathcal{T}}_{\text{quantity}}{}^{\nu} \overbrace{}^{\mu}{}^{\text{flow direction}} .$$

For example:

$$\underbrace{\mathcal{T}}_{\text{energy density}}{}^{0} \overbrace{0}^{\text{flow in time direction}} .$$

or

$$\underbrace{\mathcal{T}}_{\text{energy density}}{}^{0} \overbrace{1}^{\text{flow in } x\text{-direction}} .$$

or

$$\underbrace{\mathcal{T}}_{x\text{-momentum density}}{}^{1} \overbrace{0}^{\text{flow in time direction}} .$$

or

$$\underbrace{\mathcal{T}}_{x\text{-momentum density}}{}^{1} \overbrace{1}^{\text{flow in } x\text{-direction}} .$$

or

$$\underbrace{\mathcal{T}}_{x\text{-momentum density}}{}^{1} \overbrace{2}^{\text{flow in } y\text{-direction}} .$$

The last two examples correspond to normal stress (or pressure) and to shear stress, respectively. Overall we have

$$T^{\nu\mu} = \begin{pmatrix} T^{00} & T^{01} & T^{02} & T^{03} \\ T^{10} & T^{11} & T^{12} & T^{13} \\ T^{20} & T^{21} & T^{22} & T^{23} \\ T^{30} & T^{31} & T^{32} & T^{33} \end{pmatrix} . \tag{4.23}$$

The whole thing is a tensor—even though this is not yet clear. We can argue that it is a physical thing. So it better be a tensor. It is called the energy-momentum tensor.

For the energy itself we write the continuity equation

$$\frac{\partial T^{0\mu}}{\partial x^\mu} = 0 . \tag{4.24}$$

Now we also do this for the next line, i.e.

$$\frac{\partial T^{1\mu}}{\partial x^\mu} = 0 . \tag{4.25}$$

All in all

$$\frac{\partial T^{\nu\mu}}{\partial x^\mu} = 0 . \tag{4.26}$$

This is local conservation of energy in special relativity. In general relativity the equation is modified.

Which of the components of the energy-momentum tensor are likely to be big or small in the slow-weak field limit? What is the energy of a particle in free space: $E = m + \frac{1}{2}mv^2$. What is its momentum: $p = mv$. Note that $m \gg mv \gg mv^2$. Thus

$$\begin{pmatrix} T^{00} & \text{small} & \text{small} & \text{small} \\ \text{small} & \text{small} & \text{small} & \text{small} \\ \text{small} & \text{small} & \text{small} & \text{small} \\ \text{small} & \text{small} & \text{small} & \text{small} \end{pmatrix} = \begin{pmatrix} \rho & 0 & 0 & 0 \\ 0 & 0 & 0 & 0 \\ 0 & 0 & 0 & 0 \\ 0 & 0 & 0 & 0 \end{pmatrix} + \text{small} , \tag{4.27}$$

where ρ is the energy density. What else do we have? Remember (2.48), i.e.

$$\Delta_{\vec{r}}\, \varphi(\vec{r}) = 4\pi G \rho(\vec{r}) .$$

And we have (2.66), i.e.

$$g_{00} \approx 1 - \frac{2GM}{r} = 1 + 2\varphi .$$

Thus

$$\vec{\nabla}^2 g_{00} = 8\pi G T_{00} .$$ (4.28)

This is not a tensor equation—at least it is not clear.[2] It probably is not true in an arbitrary frame of reference. But is there a tensor equation, which in the slow weak-field limit becomes equal to this?

4.3 Einstein's Field Equations

We expect an equation like the following:

$$\mathcal{G}_{\mu\nu} = 8\pi G T_{\mu\nu} .$$ (4.29)

We do not yet know what $\mathcal{G}_{\mu\nu}$ is, but if it has (4.28) as a limiting case, it should contain second derivatives of the metric. The only candidates at hand are

$$g_{\mu\nu} \mathcal{R} \quad \text{or} \quad \mathcal{R}_{\mu\nu} .$$

so we can guess that

$$A\, g_{\mu\nu} \mathcal{R} + B\, \mathcal{R}_{\mu\nu} = 8\pi G T_{\mu\nu} ,$$ (4.30)

where A and B are constants.

Remember the previous conservation laws

$$\partial_\mu j^\mu = 0 \quad \text{and} \quad \partial_\mu T^{\nu\mu} = 0 ?$$ (4.31)

Note that this form does not really make sense, because it is not a covariant derivative! Correct, i.e. true in every reference frame, is

$$\nabla_\mu T^{\nu\mu} = 0 .$$ (4.32)

Note that this does not mean total energy and momentum conservation. The 'geometry side' of the field equation we develop here also 'contains' energy and momentum—in the form of gravitational waves.

Let's try this on (4.30). Remembering (4.17) we do see that we can make this work if $A = -1/2$ and $B = 1$! With this we obtain Einstein's field equations:

$$\boxed{\mathcal{R}_{\mu\nu} - \frac{1}{2} g_{\mu\nu} \mathcal{R} = 8\pi G T_{\mu\nu}} .$$ (4.33)

---

[2]Note that we have tacitly toggled between T^{00} and T_{00}, which here is OK. Generally, however, something like this will lead to the wrong result (cf. the example in Sect. 4.4)!

Obviously the form (4.29) is also allowed. $\mathcal{G}$ is called the Einstein tensor. There are 16 field equations, which are not all independent of course. Note that the tensors involved are symmetric, which cuts this number down to 10. Finally the Bianchi identities (4.17) reduce it to 6.

We can rewrite (4.33) into a different form, which sometimes is very useful. Contracting with $g^{\mu\nu}$ yields

$$\mathcal{R} = -8\pi G \underbrace{g^{\mu\nu} T_{\mu\nu}}_{=T^\mu_\mu = T} . \tag{4.34}$$

If we plug this into (4.33) again we obtain

$$\mathcal{R}_{\mu\nu} = 8\pi G \left(T_{\mu\nu} - \frac{1}{2} g_{\mu\nu} T \right) . \tag{4.35}$$

We see for instance that in the absence of sources, i.e. $T_{\mu\nu} = 0$, the field equations reduce to

$$\boxed{\mathcal{R}_{\mu\nu} = 0} . \tag{4.36}$$

Note that here we have used the condition $\nabla_\mu \mathcal{G}^{\nu\mu} = 0$. This does not uniquely define the Einstein tensor. Since $\nabla_\mu g^{\nu\mu} = 0$ (cf. (3.52)), a valid variation of the field equations is

$$\mathcal{G}_{\mu\nu} - \Lambda g_{\mu\nu} = 8\pi G T_{\mu\nu} , \tag{4.37}$$

where Λ is the so called cosmological constant. Note that we can move Λ to the other side of the field equations. Thus, even if $T_{\mu\nu} = 0$, which means vacuum, there can still be $\Lambda \neq 0$—which means 'vacuum energy'.

As its name is suggesting the cosmological constant is important in cosmology and we shall talk about it then. Nevertheless, we want to give an idea of its effect by considering again the Newtonian limit, i.e.

$$\vec{\nabla}^2 \varphi - \frac{1}{2}\Lambda = \underbrace{4\pi G \rho}_{=0 \text{ for the moment}} \tag{4.38}$$

We find the solution

$$\varphi = \frac{1}{12}\Lambda(x^2 + y^2 + z^2) = \frac{1}{12}\Lambda r^2 , \tag{4.39}$$

corresponding to a force linear in r. Depending on the sign of Λ the force will be repulsive or attractive. Since there is no 'center', this just means that everything is pushed away from everything else or the reverse.[3]

4.4 Weak Gravitational Fields—The Linear Limit

The above field equations are non-linear partial differential equations in the metric components. This makes them difficult to deal with. However, in many situations gravitational effects are tiny perturbations of flat spacetime. In these cases it is sensible to divide the metric tensor into two parts

$$g_{\alpha\beta} = \eta_{\alpha\beta} + h_{\alpha\beta} \,, \tag{4.40}$$

where $\eta_{\alpha\beta}$ is given by (2.70), which is just flat space time. The second term in (4.40) is small in the sense that all $|h_{\alpha\beta}| \ll 1$. We want to write down the left hand side of the field equations to leading order in the $h_{\alpha\beta}$.

It is convenient to use (4.7) to first obtain the Riemann tensor in this limit. Neglecting terms $\Gamma\Gamma \sim \mathcal{O}(h^2)$ and using the interchangeability of the partial derivatives we obtain

$$\boxed{R^{\mu}_{\ \tau\nu\sigma} = \frac{\eta^{\mu\gamma}}{2}\left(h_{\sigma\gamma,\nu\tau} - h_{\sigma\tau,\nu\gamma} - h_{\nu\gamma,\sigma\tau} + h_{\nu\tau,\sigma\gamma}\right)} \tag{4.41}$$

to first order in h. Contracting μ with ν yields the Ricci tensor to the same order in the $h_{\alpha\beta}$, i.e.

$$\begin{aligned} R_{\tau\sigma} = R^{\nu}_{\ \tau\nu\sigma} &= \frac{1}{2}\left(h_{\sigma\gamma,}{}^{\gamma}{}_{\tau} - h_{\sigma\tau,\nu}{}^{\nu}\right. \\ &\left. -\eta^{\nu\gamma}h_{\nu\gamma,\sigma\tau} + h_{\nu\tau,\sigma}{}^{\nu}\right) + \mathcal{O}(h^2) \end{aligned} \tag{4.42}$$

Finally we obtain the Ricci scalar

$$\begin{aligned} R = \eta^{\tau\sigma}R_{\tau\sigma} &= \frac{1}{2}\left(h^{\tau}_{\ \gamma,\tau}{}^{\gamma} - h^{\tau}_{\ \tau,}{}^{\nu}{}_{\nu}\right. \\ &\left. -\eta^{\nu\gamma}h_{\nu\gamma,}{}^{\tau}{}_{\tau} + h_{\nu\tau,}{}^{\nu\tau}\right) + \mathcal{O}(h^2) \,. \end{aligned} \tag{4.43}$$

At this point we combine everything into the linearised Einstein tensor

---

[3]As we shall see in the next section, the correct Newtonian limit is given by (4.51), which for $T_{\mu\nu} = 0$ implies $\vec{\nabla}^2\varphi = -\Lambda$. Our naive estimate is therefore off by a factor -2 and the potential really is $\varphi = -\frac{1}{6}\Lambda r^2$.

$$\mathcal{G}_{\tau\sigma} = \mathcal{R}_{\tau\sigma} - \frac{1}{2}\eta_{\tau\sigma}\mathcal{R}$$

$$= -\frac{1}{2}\big(\bar{h}_{\tau\sigma,\mu}{}^{\mu} + \eta_{\tau\sigma}\bar{h}_{\mu\nu}{}^{,\mu\nu} \tag{4.44}$$

$$- \bar{h}_{\tau\mu,\sigma}{}^{\mu} - \bar{h}_{\sigma\mu,\tau}{}^{\mu}\big) + \mathcal{O}(h^2)$$

where

$$\boxed{\bar{h}_{\alpha\beta} = h_{\alpha\beta} - \frac{1}{2}\eta_{\alpha\beta}h} \tag{4.45}$$

and $h = \eta^{\alpha\beta}h_{\alpha\beta}$.[4] The special appeal of this expression for $\mathcal{G}_{\tau\sigma}$ lies in the following equation, the so called Lorenz gauge[5]:

$$\boxed{\bar{h}_{\mu\nu,}{}^{\nu} = 0}. \tag{4.46}$$

Momentarily we accept this equation and explore its consequences. If we apply it to (4.44) we note that only the first term in brackets remains, i.e.

$$\mathcal{G}_{\tau\sigma} = -\frac{1}{2}\bar{h}_{\tau\sigma,\mu}{}^{\mu} = -\frac{1}{2}\Big(\frac{\partial^2}{\partial t^2} - \vec{\nabla}^2\Big)\bar{h}_{\tau\sigma}. \tag{4.47}$$

Consequently we obtain the linearised field equations:

$$\boxed{\Big(\frac{\partial^2}{\partial t^2} - \vec{\nabla}^2\Big)\bar{h}_{\tau\sigma} = -16\pi G\, T_{\tau\sigma}}. \tag{4.48}$$

Taking the trace and plugging the resulting identity into (4.45), we obtain

$$\Big(\frac{\partial^2}{\partial t^2} - \vec{\nabla}^2\Big)h_{\tau\sigma} = -16\pi G\Big(T_{\tau\sigma} - \frac{1}{2}T\eta_{\tau\sigma}\Big). \tag{4.49}$$

Let's look at this in the static limit of Newtonian gravity. Here $h_{00} = 2\varphi$ (cf. (2.68)) and thus

$$\vec{\nabla}^2\varphi = 8\pi G\Big(T_{00} - \frac{1}{2}T\Big) \tag{4.50}$$

or, more generally if we add in a cosmological constant[6]

---

[4]Note: $\bar{h} = \eta^{\alpha\beta}\bar{h}_{\alpha\beta} = -h$.

[5]Lorenz, Ludvig Valentin, Danish physicist, *Helsingoer 18.1.1829, †Frederiksberg 9.5.1981.

[6]This is the proper generalisation of the Poisson equation (2.48) that is valid when $T_{00} = \rho$ is the only non-negligible element of the energy-momentum tensor. In the special case of a homogenous

$$\boxed{\vec{\nabla}^2\varphi = 8\pi G\left(T_{00} - \frac{1}{2}T\right) - \Lambda} \qquad (4.51)$$

But how do we know that the Lorenz gauge is a valid equation?[7] Note the apparent similarity between the linearised theory and electrodynamics. Remember that in electrodynamics [9] we had defined the electromagnetic field tensor via

$$F_{\alpha\beta} = A_{\beta,\alpha} - A_{\alpha,\beta} \qquad (4.52)$$

determining the equation of motion of a charge, i.e.

$$m\frac{du_\alpha}{d\tau} = eF_{\alpha\beta}u^\beta , \qquad (4.53)$$

where u^β is the four velocity. In electrodynamics the four-potential A_α is not unique, i.e. we can apply the transformation

$$A_\alpha^{(new)} = A_\alpha^{(old)} - f_{,\alpha} , \qquad (4.54)$$

where f_α is an arbitrary function of the coordinates. This gauge transformation does not alter the electric or magnetic field. In fact, inserting (4.54) into (4.52) we see that it does not alter $F_{\alpha\beta}$.

The above mentioned similarity is between the the Riemann tensor and the electromagnetic tensor, i.e.

$$\mathcal{R}_{\alpha\beta\gamma\delta} \leftrightarrow F_{\alpha\beta} , \qquad (4.55)$$

and the $h_{\mu\nu}$ and the four-potential, i.e.

$$h_{\alpha\beta} \leftrightarrow A_\alpha . \qquad (4.56)$$

Let's assume we generalise (4.54) in the following straightforward manner,

$$h_{\alpha\beta}^{(new)} = h_{\alpha\beta}^{(old)} - \xi_{\beta,\alpha} - \xi_{\alpha,\beta} , \qquad (4.57)$$

where

$$x_\sigma' = x_\sigma + \xi_\sigma(x_\rho) . \qquad (4.58)$$

This means that we change the coordinates x_σ just slightly, i.e. (4.40) remains valid. It is not very difficult to insert (4.57) into the approximate Riemann tensor (4.41)

---

medium of density ρ and pressure P we have $T_{00} = \rho$ and $T_{ii} = P$ so that $\Delta\varphi = 4\pi G(\rho + 3P) - \Lambda$. This will be a useful relation in cosmology.

[7] The following discussion from here to the next example may be omitted in a first reading.

and find that it does not change at all. This is the equivalent of the statement that
(4.54) does not alter the electromagnetic field tensor. Because the Riemann tensor is
really our central quantity determining the motion of masses, we can use the gauge
transformation (4.57) freely and to our advantage.

From (4.57) it follows directly that

$$\bar{h}_{\alpha\beta}^{(new)} = \bar{h}_{\alpha\beta}^{(old)} - \xi_{\beta,\alpha} - \xi_{\alpha,\beta} + \eta_{\alpha\beta}\xi^{\mu}{}_{,\mu} . \tag{4.59}$$

To see this, we first note that

$$\underbrace{\eta^{\beta\alpha}h_{\alpha\beta}^{(new)}}_{=h^{(new)}} = \underbrace{\eta^{\beta\alpha}h_{\alpha\beta}^{(old)}}_{=h^{(old)}} - \underbrace{\eta^{\beta\alpha}\xi_{\beta,\alpha}}_{=\xi^{\alpha}{}_{,\alpha}} - \underbrace{\eta^{\beta\alpha}\xi_{\alpha,\beta}}_{=\xi^{\beta}{}_{,\beta}} . \tag{4.60}$$

Therefore

$$-\frac{1}{2}\eta^{\alpha\beta}h^{(new)} = -\frac{1}{2}\eta^{\alpha\beta}h^{(old)} + \eta^{\alpha\beta}\xi^{\mu}{}_{,\mu} . \tag{4.61}$$

In conjunction with the definition (4.45) we find (4.59). And from (4.59) we have

$$\bar{h}_{\alpha\beta}^{(new),\beta} = \bar{h}_{\alpha\beta}^{(old),\beta} - \xi_{\alpha,\beta}{}^{\beta} \underbrace{-\xi_{\beta,\alpha}{}^{\beta} + \eta_{\alpha\beta}\xi^{\mu}{}_{,\mu}{}^{\beta}}_{=0} . \tag{4.62}$$

Now assume that we have $h_{\alpha\beta}^{(old)} \neq 0$ but we want $h_{\alpha\beta}^{(new)} = 0$. Thus we must select
the ξ^{β} so that

$$0 = \bar{h}_{\alpha\beta}^{(old),\beta} - \xi_{\alpha,\beta}{}^{\beta} \tag{4.63}$$

is satisfied. This is always possible an indeed it is not even unique. Suppose that we
have a ξ^{β} that fulfills (4.63), we can still add to it an arbitrary ζ^{β} which fulfills

$$\zeta_{\alpha,\beta}{}^{\beta} = 0 \tag{4.64}$$

and the new $\xi'^{\beta} = \xi^{\beta} + \zeta^{\beta}$ still fulfills (4.63). We can use these remaining four
degrees of freedom to e.g. impose the conditions

$$\bar{h}_{\alpha}^{\alpha} = 0 \qquad \bar{h}_{i0} = 0 \tag{4.65}$$

which is the so called transverse traceless or TT gauge.

● **Example - Rotating Source:** The figure shows a planet of mass m revolving
around its Sun on a circular orbit in the ecliptic plane, which here is the x-

y-plane. The Sun, a spherical body of uniform density ρ, radius R, and mass M, rotates about the z-axis with a constant angular velocity $\vec{\Omega}$ and constant angular momentum $\vec{L}$. Both senses of rotation, that of the planet on its circular path around the Sun and that of the Sun itself, are the same.

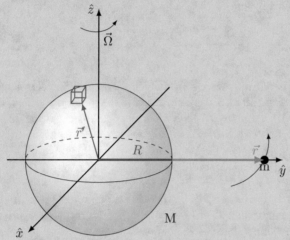

In Newtonian gravity the rotation of the Sun does not affect the field felt by the planet (here a point mass). However, in general relativity all components of $T^{\mu\nu}$ generate a field.

(a)   We begin our analysis of this system by writing down the components $T^{0\nu}$ in a Lorentz frame at rest with respect to the center of mass of the Sun, assuming ρ, Ω and R are independent of time. For every component we work to lowest non-vanishing order in ΩR.

The velocity of a volume element inside the Sun in spherical coordinates is given by

$$\dot{\vec{r}}' = \Omega \begin{pmatrix} -y' \\ x' \\ 0 \end{pmatrix} . \tag{4.66}$$

This means that the components of the energy-momentum tensor are

$$T^{\mu\nu} \approx \begin{pmatrix} \rho & -\rho\Omega y' & \rho\Omega x' & 0 \\ -\rho\Omega y' & 0 & 0 & 0 \\ \rho\Omega x' & 0 & 0 & 0 \\ 0 & 0 & 0 & 0 \end{pmatrix} . \tag{4.67}$$

Note that due to the symmetry of the energy-momentum tensor $T^{0j} = T^{j0}$. The other components T^{ij} ($i, j = 1, 3$) are $\mathcal{O}(\rho \Omega^2 R^2)$. Because $c = 1$ and thus $\Omega R \ll 1$, this means that they may be neglected here.

(b)   The general solution to the equation $\nabla^2 f = g$, vanishing at infinity, is

$$f(\vec{r}) = -\frac{1}{4\pi} \int \frac{g(\vec{r}\,')}{|\vec{r} - \vec{r}\,'|} d^3 r' . \tag{4.68}$$

We use this to solve (4.48) for $\bar{h}_{00}$ (which we actually know exactly) and $\bar{h}_{0j}$ for the source in part a). We are interested in the solution outside the Sun only and only to lowest non-vanishing order in r^{-1}, where r is the distance from the Sun's center. Expressing the result for $\bar{h}_{0j}$ in terms of the Sun's angular momentum $\vec{L}$, we obtain the metric tensor, $g_{\alpha\beta}$, within this approximation and transform it to spherical coordinates.

Hints: We may assume $\vec{r} = \vec{n}r$, where $\vec{n}$ is a unit vector, and $r \gg R$. This means one can expand $|\vec{r} - \vec{r}\,'|^{-1} \approx r^{-1}(1 + \mathcal{O}(r^{-1}))$. Note that we need the term $\mathcal{O}(r^{-1})$ but not the subsequent ones.

Since there is no explicit time dependence we have

$$\vec{\nabla}^2 \bar{h}_{\mu\nu} = 16\pi G T_{\mu\nu} . \tag{4.69}$$

Using $T_{\mu\nu} = g_{\mu\alpha} g_{\nu\beta} T^{\alpha\beta} \approx \eta_{\mu\alpha} \eta_{\nu\beta} T^{\alpha\beta}$ yields

$$T_{\mu\nu} \approx \begin{pmatrix} \rho & \rho\Omega y' & -\rho\Omega x' & 0 \\ \rho\Omega y' & 0 & 0 & 0 \\ -\rho\Omega x' & 0 & 0 & 0 \\ 0 & 0 & 0 & 0 \end{pmatrix} . \tag{4.70}$$

The solution for the 00-component is simply

$$\bar{h}_{00}(\vec{r}) = -\frac{4GM}{r} . \tag{4.71}$$

To find the other components we use

$$\frac{1}{|\vec{r} - \vec{r}\,'|} = \frac{1}{r}\left(1 + \frac{\vec{r} \cdot \vec{r}\,'}{r^2} + \mathcal{O}(R^2/r^2)\right) . \tag{4.72}$$

Note that

$$\int_{\text{sphere R}} d^3r' y' = 0$$

$$\int_{\text{sphere R}} d^3r' y' \frac{x\,x'}{r^2} = 0$$

$$\int_{\text{sphere R}} d^3r' y' \frac{y\,y'}{r^2} = \frac{y}{r^2}\frac{4\pi R^5}{15}$$

$$\int_{\text{sphere R}} d^3r' y' \frac{z\,z'}{r^2} = 0 \,.$$

Hence

$$\bar{h}_{01}(\vec{r}) \approx -\frac{16\pi R^5}{15}G\rho\,\Omega\frac{y}{r^3} = -\frac{4}{5}GMR^2\Omega\frac{y}{r^3} \,. \tag{4.73}$$

Note that the angular momentum of the Sun is

$$\vec{L} = \frac{M}{4\pi R^3/3}\int_{\text{Sun}} d^3r\,\vec{r}\times\vec{p} = \begin{pmatrix} 0 \\ 0 \\ \frac{2}{5}MR^2\Omega \end{pmatrix} \,. \tag{4.74}$$

This yields

$$\bar{h}_{01}(\vec{r}) \approx -2GL\frac{y}{r^3} \quad (=\bar{h}_{10}) \,. \tag{4.75}$$

Analogously we obtain

$$\bar{h}_{02}(\vec{r}) \approx 2GL\frac{x}{r^3} \quad (=\bar{h}_{20}) \tag{4.76}$$

and

$$\bar{h}_{03}(\vec{r}) = 0 \quad (=\bar{h}_{30}) \,. \tag{4.77}$$

The other components $\bar{h}_{\mu\nu}$ are zero to first order in ΩR.
    Next come the $h_{\mu\nu}$ using (4.45). First we have

$$\bar{h} \approx \bar{h}_{00} = -\frac{2X}{r} \,, \tag{4.78}$$

where $X = 2GM$. Thus

$$h_{\mu\nu} \approx \begin{pmatrix} -\frac{X}{r} & -\frac{Yy}{r^3} & \frac{Yx}{r^3} & 0 \\ -\frac{Yy}{r^3} & -\frac{X}{r} & 0 & 0 \\ \frac{Yx}{r^3} & 0 & -\frac{X}{r} & 0 \\ 0 & 0 & 0 & -\frac{X}{r} \end{pmatrix}, \tag{4.79}$$

where $Y = 2GL$,

$$g_{\mu\nu} \approx \begin{pmatrix} 1 - \frac{X}{r} & -\frac{Yy}{r^3} & \frac{Yx}{r^3} & 0 \\ -\frac{Yy}{r^3} & -1 - \frac{X}{r} & 0 & 0 \\ \frac{Yx}{r^3} & 0 & -1 - \frac{X}{r} & 0 \\ 0 & 0 & 0 & -1 - \frac{X}{r} \end{pmatrix}$$

and, to leading order in GM and GL,

$$g^{\mu\nu} \approx \begin{pmatrix} 1 + \frac{X}{r} & -\frac{Yy}{r^3} & \frac{Yx}{r^3} & 0 \\ -\frac{Yy}{r^3} & -1 + \frac{X}{r} & 0 & 0 \\ \frac{Yx}{r^3} & 0 & -1 + \frac{X}{r} & 0 \\ 0 & 0 & 0 & -1 + \frac{X}{r} \end{pmatrix}.$$

Using (3.19), with $x^\nu = (t, x, y, z)$ and $\partial/\partial y^\alpha = (\partial_t, \partial_r, \partial_\phi, \partial_\theta)$, we find

$$g_{\alpha\beta} \approx \begin{pmatrix} 1 - \frac{X}{r} & 0 & \frac{Y}{r} & 0 \\ 0 & -1 - \frac{X}{r} & 0 & 0 \\ \frac{Y}{r} & 0 & -(1 + \frac{X}{r})r^2 & 0 \\ 0 & 0 & 0 & -(1 + \frac{X}{r})r^2 \end{pmatrix}$$

and

$$g^{\alpha\beta} \approx \begin{pmatrix} 1 + \frac{X}{r} & 0 & \frac{Y}{r^3} & 0 \\ 0 & -1 + \frac{X}{r} & 0 & 0 \\ \frac{Y}{r^3} & 0 & -(1 - \frac{X}{r})\frac{1}{r^2} & 0 \\ 0 & 0 & 0 & -(1 - \frac{X}{r})\frac{1}{r^2} \end{pmatrix}.$$

Note that we have used $\theta = \pi/2$.

(c)   The metric is independent of t and the azimuthal angle ϕ. This implies (cf. (3.39)) that the four-momentum components $p_0 = m(1 - GM/r + J^2/2m^2r^2)$ (*) and $p_\phi = -J$ (note: we can use $J = m(GMr)^{1/2}$ from Newtonian theory) (*), are constants. (*): To $O(X^2)$ this is consistent with $p_\mu p^\mu = g^{\mu\nu} p_\nu p_\mu = m^2$.

One orbit of the planet, $\Delta\phi = 2\pi$, will take the time $T = (dt/d\tau)(d\tau/d\phi)$ $\Delta\phi$. Note that $dt/d\tau = u^0$, the zero-component of the four velocity, and $d\phi/d\tau = u^\phi$ (note: $u^0/u^\phi = p^0/p^\phi$). This can be expressed in terms of p_0, J, and the metric. Based on this we can compute Δt, the deviation from the classical time per revolution, to leading order, i.e. $\Delta t = ...L/M$.

We start from

$$\frac{dt}{d\phi} = \frac{p^0}{p^\phi} = \frac{g^{0\alpha}p_\alpha}{g^{\phi\beta}p_\beta} = \frac{g^{00}p_0 + g^{0\phi}p_\phi}{g^{\phi 0}p_0 + g^{\phi\phi}p_\phi}$$
$$\approx \frac{(1 + \frac{2GM}{r})p_0 + \frac{2GL}{r^3}p_\phi}{\frac{2GL}{r^3}p_0 - (1 - \frac{2GM}{r})\frac{1}{r^2}p_\phi}, \tag{4.80}$$

Note that we can write

$$p_0 \approx \left(1 - \frac{GM}{2r}\right)m . \tag{4.81}$$

Therefore the leading contribution to $dt/d\phi$ is

$$\frac{dt}{d\phi} \approx \frac{m}{\frac{2GL}{r^3}m - \frac{1}{r^2}p_\phi}$$
$$\approx \frac{mr^2}{J}\left(1 - \frac{2GL}{r^3}\frac{mr^2}{J}\right)$$
$$= \frac{1}{\omega}\left(1 - \frac{2GL}{r^3}\frac{1}{\omega}\right)$$
$$= \frac{1}{\omega} - 2\frac{L}{M}, \tag{4.82}$$

where we have used the classical formulas $J = mr^2\omega$ and $r^3\omega^2 = GM$. Hence

$$T \approx T_{\text{Newton}} - 4\pi\frac{L}{M} . \tag{4.83}$$

(d) Finally we want to try out some numbers. Based on our Sun ($\Omega = 3 \cdot 10^{-6}$ s^{-1}) and the Earth what is Δt per year in seconds? What would be the result if the Sun's sense of rotation was opposite?

In MKS units the last term in (4.83) is given by

$$\Delta t = -4\pi\frac{L}{Mc^2} = -\frac{4\pi}{c^2}\frac{2}{5}R_\odot^2\Omega \approx -82 \cdot 10^{-6} \text{ s} . \tag{4.84}$$

Changing the sense of rotation results in a positive Δt of the same magnitude.

4.5 Problems

1. Angular deficit: show the validity of (4.2), i.e. $\delta\theta = \delta a / R_s^2$.
2. Prove the Bianchi identity (4.13).
3. Symmetries of the Riemann tensor

   (a) How many independent components does the Riemann curvature tensor $\mathcal{R}_{\mu\nu\alpha\beta}$ have in two dimensions? How are they related to the curvature scalar?
   (b) Show that in $3 + 1$ dimensions the cyclicity property of the Riemann tensor (4.10) follows from its other symmetries and the one additional condition that its totally antisymmetric part $\varepsilon^{\mu\nu\alpha\beta} R_{\mu\nu\alpha\beta}$ vanishes. Show that in $2 + 1$ dimensions, cyclicity is implied by the other symmetries.
   (c) Use the symmetries of the Riemann tensor to determine how many *independent* components it has in $3 + 1$ and $2 + 1$ dimensions.

4. Does the Einstein equation (in $3 + 1$ dimensions) permit nontrivial vacuum solutions, i.e. solutions where the energy momentum tensor vanishes but the curvature does not? How about $2 + 1$ and $1 + 1$ dimensions? *Hint: think of the number of degrees of freedom the Riemann tensor has and if they are all determined by the energy momentum tensor.*
5. Compute the Riemann curvature tensor $\mathcal{R}_{\mu\nu\alpha\beta}$, the Ricci tensor $\mathcal{R}_{\mu\nu}$ and the curvature scalar of the surface of a sphere with radius r.
6. Imagine flat, Euclidean space that is rescaled by some time dependent factor $a(t)$. The corresponding spacetime is characterised by

$$d\tau^2 = dt^2 - a^2(t)(dx^2 + dy^2 + dz^2)$$

   (a) Determine the metric, the inverse metric and the Christoffel symbols.
   (b) Determine the nonzero elements of the Riemann curvature tensor $\mathcal{R}_{\alpha\beta\gamma\delta}$. *Hint: Since there are a lot of components in principle, it is typically a good strategy to first determine all nonzero derivatives of Christoffel symbols. Then it is typically a good idea to start from the Christoffel symbols and their derivatives and build nonzero elements of the Riemann tensor. Once you have such an element, use all of the Riemann tensors symmetries to obtain as many other elements as possible in a trivial way. In the current example, there are really just two independent elements.*
   (c) Contract the Riemann tensor and obtain the Ricci tensor $\mathcal{R}_{\mu\nu}$ and scalar $\mathcal{R}$.
   (d) How is the energy-momentum tensor $T_{\mu\nu}$ that produces such a metric related to the scale factor a? If we want the scale factor to be time independent, what does this imply for $T_{\mu\nu}$?

7. Energy-momentum tensor of a perfect non-relativistic fluid: Consider a non-relativistic perfect fluid for which $u^0 \approx 1$, $u^i \ll 1$ ($i = 1, 2, 3$) and $P \ll \rho$. Here u^α is the four-velocity, P is the pressure and ρ is the energy density. Show that in flat spacetime $\partial T^{\alpha\beta}/\partial x^\beta = 0$ is equivalent to mass conservation plus

Euler's equation of motion. Hints: The energy-momentum tensor for a perfect relativistic fluid is given by $T_\alpha^\beta = (\rho + P)u_\alpha u^\beta - P\delta_\alpha^\beta$; Euler's equation is $\rho\, d\vec{v}/dt + \vec{\nabla}P = 0$ (cf. L. D. Landau and E. M. Lifshitz *Fluid Mechanics*. Elsevier 2009).

Chapter 5
Classical Tests of General Relativity

The most iconic experimental tests of general relativity for weak gravitational fields are presented. Those include the classic 20th century observations of the the perihelion precession and the deflection of light by massive objects such as the sun as well as the 21st century detection of gravitational waves.

5.1 Schwarzschild Solution

In vacuum $T_{\nu\mu}$ vanishes and we have (4.36), i.e.

$$\mathcal{R}_{\nu\mu} = 0 \ .$$

We seek a spherically symmetric, time-independent solution corresponding to a single point mass M at the origin. Let's try the ansatz

$$d\tau^2 = e^{2n(r)}dt^2 - e^{2m(r)}dr^2 - r^2 \left(d\theta^2 + \sin^2\theta d\phi^2 \right) \ . \tag{5.1}$$

Thus

$$g_{00} = e^{2n(r)} \quad g_{11} = -e^{2m(r)} \quad g_{22} = -r^2 \quad g_{33} = -r^2 \sin^2\theta \ . \tag{5.2}$$

Using one of the two symbolic algebra routines in the appendix we obtain the following non-vanishing components of the Ricci tensor which we set to 0:

© Springer Nature Switzerland AG 2020
R. Hentschke and C. Hölbling, *A Short Course in General Relativity and Cosmology*, Undergraduate Lecture Notes in Physics,
https://doi.org/10.1007/978-3-030-46384-7_5

$$\mathcal{R}_{00} = e^{2n(r)-2m(r)} \left(n''(r) + n'(r)^2 - n'(r)m'(r) + \frac{2n'(r)}{r} \right) = 0 \qquad (5.3)$$

$$\mathcal{R}_{11} = -n''(r) - n'(r)^2 + n'(r)m'(r) + \frac{2m'(r)}{r} = 0 \qquad (5.4)$$

$$\mathcal{R}_{22} = e^{-2m(r)} \left(-rn'(r) + rm'(r) - 1 \right) + 1 = 0 \qquad (5.5)$$

$$\mathcal{R}_{33} = \mathcal{R}_{22} \sin^2 \theta = 0 \qquad (5.6)$$

Equation (5.6) is automatically true because of (5.5). The combination of (5.3) and (5.4) yields

$$n'(r) + m'(r) = 0 \qquad (5.7)$$

or

$$n(r) + m(r) = const . \qquad (5.8)$$

In the limit $r \to \infty$ we expect the flat spacetime metric and thus ($n(r) = 0$ and $m(r) = 0$ in this limit) $const = 0$. Hence

$$n(r) + m(r) = 0 . \qquad (5.9)$$

Using (5.5) and (5.9) yields

$$e^{2n(r)} \left(2rn'(r) + 1 \right) = 1 . \qquad (5.10)$$

or

$$\frac{d}{dr} \left(re^{2n(r)} \right) = 1 . \qquad (5.11)$$

or

$$re^{2n(r)} = r - 2GM , \qquad (5.12)$$

where $-2GM$ is the constant of integration. We find the Schwarzschild metric

$$\boxed{ d\tau^2 = \left(1 - \frac{2GM}{r} \right) dt^2 - \frac{dr^2}{1 - \frac{2GM}{r}} - r^2 d\Omega^2 } , \qquad (5.13)$$

where $d\Omega^2 = d\theta^2 + \sin^2 \theta d\phi^2$. 'It can be shown' that to within unimportant coordinate transformations this is the only metric which obeys $\mathcal{R}_{\mu\nu} = 0$ ($r \neq 0$) and (1) is independent of t, (2) depends spatially only on r, and (3) goes over into flat spacetime for $r \to \infty$. Recall (cf. (2.66)) that in the classical limit

$$g_{00} = 1 - \frac{2GM}{r}$$

Therefore M is interpreted as point mass at $r = 0$.

5.2 Geodesics in the Schwarzschild Metric

Let us compute geodesics in the Schwarzschild metric to see how small objects move under the influence of a large mass. The largest mass we have in our vicinity is the Sun and its Schwarzschild radius $r_s = 2GM \sim 2.95$ km is far smaller than its radius. We are therefore interested in geodesics for which $r_s \ll r$. In addition, we choose the coordinates such that the motion takes place in the $\theta = \pi/2$ plane, which will simplify our expressions. In this plane, the metric is

$$g_{\mu\nu} = \begin{pmatrix} 1 - \frac{r_s}{r} & 0 & 0 & 0 \\ 0 & -\frac{1}{1-\frac{r_s}{r}} & 0 & 0 \\ 0 & 0 & -r^2 & 0 \\ 0 & 0 & 0 & -r^2 \end{pmatrix} \tag{5.14}$$

The equations of motion (2.58) with a parameter p describing the path are

$$\frac{d^2x^\mu}{dp^2} + \Gamma^\mu_{\alpha\beta} \frac{dx^\alpha}{dp} \frac{dx^\beta}{dp} = 0 . \tag{5.15}$$

We use (2.59) (or the symbolic algebra routines in Appendix F) to work out the Christoffel symbols. The resulting equations are

$$\frac{d^2t}{dp^2} = -\frac{r_s}{r(r - r_s)} \frac{dr}{dp} \frac{dt}{dp} \tag{5.16}$$

$$\frac{d^2r}{dp^2} = -\frac{r_s(r - r_s)}{2r^3} \left(\frac{dt}{dp}\right)^2 + \frac{r_s}{2r(r - r_s)} \left(\frac{dr}{dp}\right)^2 + (r - r_s) \left(\frac{d\phi}{dp}\right)^2 \tag{5.17}$$

$$\frac{d^2\phi}{dp^2} = -\frac{2}{r} \frac{dr}{dp} \frac{d\phi}{dp} \tag{5.18}$$

Note first that (5.16) and (5.18) can be rewritten as

$$\frac{d}{dp} \left(\frac{r - r_s}{r} \frac{dt}{dp}\right) = 0 \tag{5.19}$$

and

$$\frac{d}{dp}\left(r^2\frac{d\phi}{dp}\right) = 0 \ . \tag{5.20}$$

Equation (5.19) can be used to tie the parameter p to the coordinate time t via

$$\frac{r-r_s}{r}\frac{dt}{dp} = H \ , \tag{5.21}$$

where H is a constant of motion. Equation (5.20) yields a second constant of motion, J, i.e.

$$r^2\frac{d\phi}{dp} = J \ . \tag{5.22}$$

Inserting (5.21) and (5.22) into (5.17) yields

$$\frac{d^2r}{dp^2} - \frac{r-r_s}{r^4}J^2 - \frac{r_s}{2r(r-r_s)}\left(\left(\frac{dr}{dp}\right)^2 - H^2\right) = 0 \ . \tag{5.23}$$

This equation can be expressed as the d/dp-derivative of

$$\frac{r}{r-r_s}\left(\left(\frac{dr}{dp}\right)^2 - H^2\right) + \frac{J^2}{r^2} = -K \ , \tag{5.24}$$

where K is a third constant of motion. It will turn out to be convenient to use the inverse radius $u = 1/r$ instead of the radius r. If we make this substitution and use the identity

$$\frac{dr}{dp} = -r^2\frac{du}{d\phi}\frac{d\phi}{dp} = -J\frac{du}{d\phi} \ , \tag{5.25}$$

we can rewrite (5.24) as an equation for the trajectory

$$\frac{du}{d\phi} = \pm\sqrt{\frac{H^2-K}{J^2} + \frac{Kr_su}{J^2} - u^2 + r_su^3} \tag{5.26}$$

The combination of (5.21), (5.22) and (5.24) yields

$$d\tau^2 = Kdp^2 \ . \tag{5.27}$$

For photons $d\tau^2 = 0$ and thus

$$\text{photons:} \quad K = 0, \quad \frac{du}{d\phi} = \pm\sqrt{\frac{H^2}{J^2} - u^2 + r_su^3} \ , \tag{5.28}$$

whereas for massive particles we may choose

$$\text{particles:} \quad K = 1, \quad \frac{du}{d\phi} = \pm\sqrt{\frac{H^2 - 1}{J^2} + \frac{r_s u}{J^2} - u^2 + r_s u^3} \tag{5.29}$$

so that in this case we can parameterise the curve with the eigentime $\tau = p$.

5.3 Perihelion Precession

Mercury, like all celestial bodies in the solar system, does not orbit in a perfect ellipse but rather in a rosetta. Most of the drift of the point of its closest approach to the Sun, the perihelion precession, is due to the influence of other planets. However, by the late 19th century it became apparent that about 8% or 43″ per century ($1'' = 4.8481 \cdot 10^{-6}$ radians) of this precession could not be accounted for by Newtonian celestial mechanics. Let's see whether Einstein's theory can do better.

In Newtonian mechanics (e.g., (5.30) in [4] on page 131) we have

$$\left(\frac{d\phi}{dr}\right)_N = \pm\frac{L}{r^2}\left[2\mu E + 2\mu\frac{GmM}{r} - \frac{L^2}{r^2}\right]^{-1/2}. \tag{5.30}$$

Here L is the angular momentum (note: $L = \mu r^2 d\phi/dt$), E is the total energy, and $\mu = mM/(m + M)$ is the reduced mass. The solution of this equation is an ellipse:

$$r = \frac{r_0}{1 + e\cos\phi}, \tag{5.31}$$

with $r_0 = L^2/(\mu\alpha)$ and the excentricity $e = \sqrt{1 + 2\frac{EL^2}{\mu\alpha^2}} < 1$ ($\alpha = mMG$). We can again substitute $u = 1/r$ (and $u_0 = 1/r_0$) to find

$$\left(\frac{du}{d\phi}\right)_N = \pm\sqrt{u_0^2(e^2 - 1) + 2u_0 u - u^2} \tag{5.32}$$

with the ellipse solution

$$u = u_0(1 + e\cos\phi). \tag{5.33}$$

We can now go back to (5.29) and substitute $u_0 = r_s/(2J^2)$ and $u_0^2(e^2 - 1) = (H^2 - 1)/J^2$ thus trading the old pair of integration constants H and J for a new pair u_0 and e that have a direct geometrical meaning in the Newtonian limit. If we make these substitutions in (5.29) we find

$$\frac{du}{d\phi} = \pm\sqrt{u_0^2(e^2 - 1) + 2u_0 u - u^2(1 - r_s u)} \qquad (5.34)$$

and we note that general relativity just adds a small correction term $r_s u = r_s/r \ll 1$ to the Newtonian trajectory. Although it is possible to solve this differential equation exactly in terms of elliptic functions, it is easier and more transparent to compute the precession of the perihelion to first order in a perturbative expansion. Inserting the ansatz

$$u(\phi) = u_0(1 + f(\phi)) \qquad (5.35)$$

into (5.34) we obtain

$$\frac{df}{d\phi} = \pm\sqrt{(e^2 - f^2) + r_s u_0(1 + f)^3} \,. \qquad (5.36)$$

Note that in the Newtonian limit, i.e. $r_s u_0 \to 0$, $f(\phi) \to e \cos\phi$ (cf.(5.33)). To obtain the perihelion precession, we integrate the trajectory from one minimum of r (maximum of u) to the next and compare the corresponding angle ϕ to 2π for a closed orbit. In fact, because of time reversal invariance of the problem, we just need to integrate the trajectory from a minimum of u to the subsequent maximum and compare the angle with π, i.e.

$$\int_{f_{min}}^{f_{max}} df \frac{1}{\sqrt{e^2 - f^2 + r_s u_o(1 + f)^3}} = \pi + \frac{\delta\phi}{2} \,. \qquad (5.37)$$

The extrema f_{max} and f_{min} follow via $\frac{df}{d\phi} = 0$. We do not need to compute them explicitly though since we solve the integral by a variable substitution $g = f\sqrt{1 - r_s u_0(1 + f)^3/f^2}$ so that $g^2 = f^2 - r_s u_0(1 + f)^3$. For this new variable the condition $\frac{df}{d\phi} = 0$ implies $g_{max/min} = \pm e$ and we obtain

$$\int_{-e}^{e} dg \frac{1}{\sqrt{e^2 - g^2}} \frac{g}{f - \frac{3}{2} r_s u_0(1 + f)^2} = \pi + \frac{\delta\phi}{2} \,. \qquad (5.38)$$

To first order in $r_s u_0$ we can rewrite this as

$$\delta\phi = 2 \underbrace{\int_{-e}^{e} \frac{dg}{\sqrt{e^2 - g^2}}}_{=\pi} + r_s u_0 \underbrace{\int_{-e}^{e} \frac{dg}{\sqrt{e^2 - g^2}} \frac{2g - 1}{g^2}(1 + g)^2 - 2\pi}_{=3\pi} \qquad (5.39)$$

$$= 3\pi r_s u_0 \,.$$

Expressing $u_0 = 1/(a(1 - e^2))$, where a is the semi-major axis of the ellipse and $r_s = 2GM$, we finally find

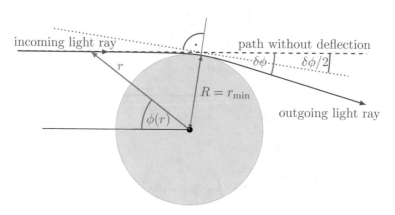

Fig. 5.1 Deflection of light passing close to a mass with radius R

$$\boxed{\delta\phi \approx 6\pi \frac{GM}{a(1-e^2)}} \, . \tag{5.40}$$

The quantity in the denominator is called semilatus rectum. In the case of Mercury it is 55.46 million km. In SI-units we also have $GM_\odot/c^2 = 1.475\,\text{km}$ and thus

$$\delta\phi_{\text{Mercury}} \approx 0.103'' \text{ per revolution} \, . \tag{5.41}$$

or

$$\delta\phi_{\text{Mercury}} \approx 43.03'' \text{ per century} \tag{5.42}$$

The precessions of other planets can also be calculated from general relativity and agree with respective measurements (cf. for instance Table 8.3 in [11]).

5.4 Deflection of Light

Mathematically this problem is quite similar to the precession problem—except that the orbit is unbound. Conceptually, however, it is quite different, because we are dealing with light and not with massive particles.

The trajectory of a light ray is described by (5.28) and the situation is depicted in Fig. 5.1. Without deflection the angle ϕ varies between 0, i.e. the light ray is approaching from a large distance, to π, i.e. the light rays disappear. With attractive deflection we expect $\pi + \delta\phi$ instead. At the minimum distance $r_{\text{min}} = R$, i.e. at the maximum inverse distance $u_0 = 1/R$ the derivative $du/d\phi|_{r_{\text{min}}} = 0$, which, according to (5.28) fixes the remaining integration constants to

$$\left(\frac{H}{J}\right)^2 = u_0^2 - r_s u_0^3 \, . \tag{5.43}$$

Equation (5.28) thus becomes

$$-\frac{du}{d\phi} = \sqrt{u_0^2 - u^2 - r_s(u_0^3 - u^3)} \, . \tag{5.44}$$

Now we integrate both sides, i.e.

$$\int_0^{(\pi+\delta\phi)/2} d\phi = -\int_{u_0}^0 \frac{du}{\sqrt{u_0^2 - u^2 - r_s(u_0^3 - u^3)}} \tag{5.45}$$

or

$$\frac{\pi}{2} + \frac{1}{2}\delta\phi = \int_0^{u_0} \frac{du}{\sqrt{u_0^2 - u^2 - r_s(u_0^3 - u^3)}} \, . \tag{5.46}$$

Here the term $r_s(u^3 - u_0^3)$ is the perturbation, i.e.

$$\frac{1}{2}\delta\phi \approx \int_0^{u_0} du \, \frac{r_s(u_0^3 - u^3)}{2(u_0^2 - u^2)^{3/2}} \, . \tag{5.47}$$

or

$$\delta\phi \approx r_s u_0 \underbrace{\int_0^1 dx \, \frac{1-x^3}{(1-x^2)^{3/2}}}_{=2} \quad (x = \frac{u}{u_0}) \, . \tag{5.48}$$

The final result is

$$\boxed{\delta\phi \approx \frac{4GM}{R}} \, . \tag{5.49}$$

For the Sun $\delta\phi \approx 1.75''$ (for Jupiter it's only $0.02''$).

Measurements (photographs) during the solar eclipse on May 29, 1919 confirmed Einstein's value and made him famous.

Remark 'Modern' versions of this experiment were carried out using radar echo delay in particular by I. I. Shapiro. This test of Einstein's theory is similar to the light-bending experiment depicted in Fig. 5.1. Here, however, it is not starlight that is approaching from the left but radio signals from an emitter on Earth. The electromagnetic signal then passes close to the Sun. Instead of being detected at some distance from the Sun on the other side, the signal is reflected by the surface of a planet, e.g. Mercury. It returns along the same path towards Earth and arrives at time t after its original departure. The presence of the Sun along the signal's path gives rise to a delay $\Delta t_{\text{round-trip}}$. Starting point is (5.24). Elimination of the parameters p and J/H leads to an equation for dt/dr, which is integrated to yield $t(r, R)$. The quantity R is the distance of closest approach to the Sun's center (basically its radius) and r is some distance measured from the trajectory to the center of the Sun (cf. Fig. (5.1)). The result is

$$t(r, R) = R\sqrt{x^2 - 1} + GM \left\{ \sqrt{\frac{x - 1}{x + 1}} + 2\ln\left(x + \sqrt{x^2 - 1}\right) \right\}, \quad (5.50)$$

where $x = r/R$. The maximum time delay for the round-trip Earth-Sun-Mercury and back, calculated with this formula, is roughly 240 μs (see also [11]; Sect. 7).

5.5 Gravitational Waves

In context of the linear theory we arrived at (4.48), which in vacuum ($T_{\tau\sigma} = 0$) becomes

$$\left(\frac{\partial^2}{\partial t^2} - \vec{\nabla}^2\right)\bar{h}_{\tau\sigma} = 0. \quad (5.51)$$

This looks just like the wave equation in electrodynamics [9] with the four-potential replaced by $\bar{h}_{\tau\sigma}$. Equation (5.51) describes weak gravitational waves. From this equation we see immediately the gravitational waves move at the same speed as the electromagnetic waves in vacuum. Just as in electrodynamics we find that

$$\bar{h}_{\tau\sigma} = A_{\tau\sigma} \exp[ik_\lambda x^\lambda] \quad (5.52)$$

is a solution of (5.51). Inserting this into (5.51) yields

$$k_\sigma k^\sigma = 0. \quad (5.53)$$

With

$$k_\sigma \rightarrow (\omega, -\vec{k}), \quad (5.54)$$

this means that

$$\omega = |\vec{k}| \quad (5.55)$$

is the attendant dispersion relation. The wave's velocity is 1 (or $c = 1$!).

Now note that from (5.52) we have

$$\bar{h}_{\tau\sigma}{}^{,\mu} = ik^{\mu}\bar{h}_{\tau\sigma} .$$ (5.56)

But we also have the gauge condition (4.46), i.e. $\bar{h}_{\mu\nu}{}^{,\nu} = 0$, and thus

$$0 = k^{\sigma}\bar{h}_{\tau\sigma} \quad \text{or} \quad k^{\sigma}A_{\tau\sigma} = 0 .$$ (5.57)

For a plane wave in say z-direction, i.e. the wave only depends on the one spatial coordinate which happens to be z, we have $k^0 \neq 0$ and $k^3 \neq 0$. We impose the TT gauge (4.65) and use the symmetry of the metric tensor to find that $A_{\tau\sigma}$ has only two independent components, A_{xx}^{TT} and A_{xy}^{TT} and can be put into the form

$$\left(A_{\tau\sigma}^{TT}\right) = \begin{pmatrix} 0 & 0 & 0 & 0 \\ 0 & A_{xx}^{TT} & A_{xy}^{TT} & 0 \\ 0 & A_{xy}^{TT} & -A_{xx}^{TT} & 0 \\ 0 & 0 & 0 & 0 \end{pmatrix} .$$ (5.58)

Let's consider two nearby particles, one at the origin and the other one at say $x = l_o$ and $y = z = 0$. Both particles are at rest in some inertial frame. What happens when the particles are hit by an incoming gravitational wave? The equation of motion is (2.58), or

$$\frac{du^{\mu}}{d\tau} + \Gamma_{\alpha\beta}^{\mu}u^{\alpha}u^{\beta} = 0 .$$ (5.59)

where $u^{\alpha} = dx^{\alpha}/d\tau$ is the particle's four-velocity. So, initially when the wave hits

$$\frac{du^{\mu}}{d\tau}\bigg|_{o} = -\Gamma_{00}^{\mu} = -\eta^{\mu\sigma}\frac{1}{2}\left(h_{\sigma 0,0}^{TT} + h_{0\sigma,0}^{TT} - h_{00,\sigma}^{TT}\right) .$$ (5.60)

Due to (5.58) $h_{\sigma 0}$ vanishes and the particle does not accelerate. This situation does not change, which means the particle stays at its coordinate. The same of course is true for the other particle. It too stays at its coordinate.

But now let's look at the proper distance of the two particles, i.e.

$$\begin{aligned} l &\equiv \int |d\tau^2|^{1/2} = \int |g_{\alpha\beta}dx^{\alpha}dx^{\beta}|^{1/2} \\ &= \int_0^{\epsilon} |g_{xx}|^{1/2}dx \approx |g_{xx}(x = 0)|^{1/2}\epsilon \\ &\approx \left(1 + \frac{1}{2}h_{xx}^{TT}(x = 0)\right)\epsilon \end{aligned}$$ (5.61)

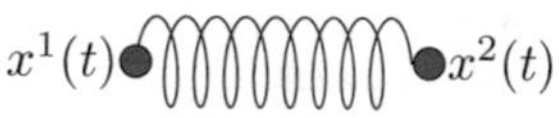

Fig. 5.2 Two identical masses m at $x^1(t)$ and $x^2(t)$ connected by a spring

$$x^1(t)\,\bullet\!\text{\scriptsize〰〰〰〰}\!\bullet\,x^2(t)$$

Because h_{xx}^{TT} is not generally zero, the proper distance does indeed change with time.

This is not as paradox as it may seem. Imagine a spherical balloon. Use a suitable pen and put two dots on its surface. The coordinates of each dot are its longitude and its latitude. Now inflate the balloon to a bigger size. Do the coordinates of the dots change? No, their longitudes and latitudes are the same. Does their distance change? Yes. We shall encounter this again when we talk about the expansion of the universe due to a time-dependent metric.

At this point let's discuss how a detector for gravitational waves could look like.[1] A cartoon of a so called resonance detector is depicted in Fig. 5.2. In classical mechanics the equations of motion of the masses are

$$x^1{}_{,00} = -\frac{1}{2}\omega_o^2(x^1 - x^2 + l_o) \tag{5.62}$$

and

$$x^2{}_{,00} = -\frac{1}{2}\omega_o^2(x^2 - x^1 - l_o)\,. \tag{5.63}$$

If we define $\xi = x^2 - x^1 - l_o$ we can combine the two equations to yield

$$\xi_{,00} + \omega_o^2\xi = 0\,. \tag{5.64}$$

This is the simple harmonic oscillator with an angular frequency ω_o. Now we assume that a gravitational wave passes this detector. The gravitational wave changes the metric, straining the spring. This results in a driving force, which can be included on the right hand side of the previous equation:

$$\boxed{\xi_{,00} + \omega_o^2\xi = \frac{1}{2}l_o h_{xx,00}^{TT}}\,. \tag{5.65}$$

In order to understand how this equation comes about let's consider the following illustrative example. Figure 5.3 shows the gravitational wave detector constructed by scientists living in a one-dimensional world. The coordinates of the two masses are θ_o^1 and θ_o^2. Their equivalent of (5.64) is

$$\Delta\theta_{o,00} + \omega_o^2\Delta\theta_o = 0\,, \tag{5.66}$$

---

[1] We recommend reading chapter 27 in *The Role of Gravitation in Physics, Report from the 1957 Chapel Hill Conference*. C. M. DeWitt and D. Rickles (eds.), where R. P. Feynmann discusses an early version of a gravitational wave detector based on (5.72).

Fig. 5.3 Gravitational wave
detector on a circle

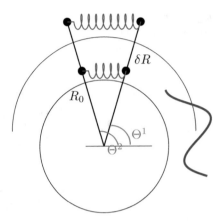

where $\Delta\theta_o = \theta_o^2 - \theta_o^1$. Note that the scientists do not perceive their world as a circle. They do not know what a circle is and they do not know what polar coordinates are. For them their world is flat and their coordinates are uniformly spaced ticks and not polar coordinates in a higher dimension.

The blue wiggle in Fig. 5.3 is a gravitational wave passing through this simple world. Its effect is to increases the world's radius by $\delta R(t)$ from R_o to $R_o + \delta R(t)$. What happens to the distance Δs between ticks? The answer is

$$\Delta s^2(t) = R^2(t)\Delta\theta^2 = (R_o + \delta R(t))^2 \Delta\theta^2 \approx R_o^2 \left(1 + 2\underbrace{\frac{\delta R(t)}{R_o}}_{=h(t)}\right)\Delta\theta^2 \quad (5.67)$$

and thus

$$\Delta s(t) \approx R_o \left(1 + \frac{1}{2}h(t)\right)\Delta\theta \quad (5.68)$$

(cf. (5.61)). The distortion $\delta\Delta s(t) = \Delta s(t) - \Delta s_o$, where $\Delta s_o = R_o \Delta\theta$, can be expressed as a correspondingly altered $\Delta\theta$, i.e. $\delta\Delta s(t) \approx R_o \delta\Delta\theta(t)$ or

$$\delta\Delta\theta(t) \approx \frac{1}{2}h(t)\Delta\theta . \quad (5.69)$$

Taking the second derivative with respect to time we find

$$\delta\Delta\theta(t)_{,00} = \Delta\theta_{,00}(t) \approx \frac{1}{2}h_{,00}(t) . \quad (5.70)$$

What's happening is that the gravitational wave stretches the spring like an external force drives an oscillator. Equation (5.70) in addition tells us that we can accommodate the external force as an extra acceleration. Thus (5.66) in the presence of the gravitational force becomes

$$\Delta\theta_{o,00} + \omega_o^2 \Delta\theta_o = \frac{1}{2}h_{,00} , \qquad (5.71)$$

We want to look at the right hand side of (5.65) in a more formal way. In Appendix B we discuss the equation for the geodesic deviation, i.e. (B.7). But the version we need here is (B.6), i.e.

$$\nabla_\tau \nabla_\tau \xi^\alpha = \mathcal{R}_{00\beta}^\alpha \xi^\beta = -\mathcal{R}_{0\beta0}^\alpha \xi^\beta . \qquad (5.72)$$

This means we take the position as observer initially at point A in Fig. B.1. Using (4.41) we find in the TT gauge

$$\mathcal{R}_{0x0}^x = \mathcal{R}_{x0x0} = -\frac{1}{2}h_{xx,00}^{TT}$$

$$\mathcal{R}_{0x0}^y = \mathcal{R}_{y0x0} = -\frac{1}{2}h_{xy,00}^{TT}$$

$$\mathcal{R}_{0y0}^y = \mathcal{R}_{y0y0} = -\frac{1}{2}h_{yy,00}^{TT} = -\mathcal{R}_{0x0}^x .$$

Now we set ξ^β in (5.72) equal to ϵ. In addition we realize that for us $\nabla_\tau \nabla_\tau \xi^\alpha = \frac{\partial^2 \xi^\alpha}{\partial t^2}$, where t is our clock time. This means that two particles initially separated by ϵ in the x-direction have a separation vector obeying

$$\frac{\partial^2 \xi^x}{\partial t^2} = \frac{1}{2}\epsilon \frac{\partial^2 h_{xx}^{TT}}{\partial t^2}$$

$$\frac{\partial^2 \xi^y}{\partial t^2} = \frac{1}{2}\epsilon \frac{\partial^2 h_{xy}^{TT}}{\partial t^2}$$

Note the consistency with (5.61)! Two particles initially separated by ϵ in the y-direction have a separation vector obeying

$$\frac{\partial^2 \xi^y}{\partial t^2} = \frac{1}{2}\epsilon \frac{\partial^2 h_{yy}^{TT}}{\partial t^2} = -\frac{1}{2}\epsilon \frac{\partial^2 h_{xx}^{TT}}{\partial t^2}$$

$$\frac{\partial^2 \xi^x}{\partial t^2} = \frac{1}{2}\epsilon \frac{\partial^2 h_{xy}^{TT}}{\partial t^2}$$

Now suppose a wave arrives which has $h_{xx}^{TT} \neq 0$ but $h_{xy}^{TT} = 0$. What happens is illustrated in Fig. 5.4. The black circle shows particles along the rim of a circle of radius $l = \epsilon$. The deformation of the circular arrangement due to the wave is illustrated by the red dots. This is one possible polarisation state. If the wave has

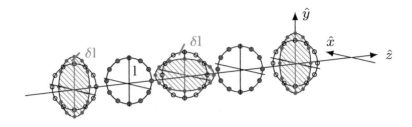

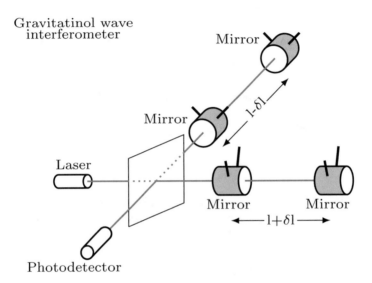

Fig. 5.4 Top: a periodic spatial deformation of a circular arrangement of test masses in the x-y-plane at different times due to a gravitational wave in z-direction. Bottom: attendant cartoon of the LIGO detector discussed below

$h_{xx}^{TT} = 0$ but $h_{xy}^{TT} \neq 0$ instead, the corresponding figure can be obtained from Fig. 5.4. by a simple rotation around the z-axis by 45°.

Let's return to our simple detector. The solution of (5.65) is rather straightforward if we assume the gravitational wave is of the simple form

$$h_{xx}^{TT} = A \cos \omega t , \qquad (5.73)$$

with a wavelength much larger than the distance between the masses. Note that we do not pay attention here to the details of the polarisation of the gravitational wave relative to the orientation of the detector. Inserting (5.73) in (5.65) results in the standard form of a driven oscillator in mechanics and we can adopt its solution from any textbook on the subject (e.g. [4]; p. 156). The amplitude of the (long-time) response, i.e. $\xi = \xi_o \cos \omega t,$[2] is given by

$$\xi_o = \frac{1}{2} l_o A \frac{\omega^2}{|\omega_o^2 - \omega^2|} . \tag{5.74}$$

What this is telling us is that we want a large detector, a large amplitude of the gravitational wave and we want its frequency in resonance with our detector.

In the 1960 Joseph Weber, an american scientist at the University of Maryland, tried to use large aluminum cylinders weighing several tons as gravitational wave detectors. To pick up the small oscillations he used piezoelectric crystals around the circumference of the cylinders. It is also important to install at least one more or better several additional detectors in laboratories separated by large distances. This helps to rule out false signals due to local sources. In 1969 Weber declared that his antenna had received gravitational wave signals. However, the strength and origin of these signals would have implied that the Milky Way had radiated off far more energy than it currently contains in total. In addition, other scientists tried to repeat Weber's experiments but without success. Obviously, the problem is that the $h_{\alpha\beta}^{TT}$ as well as their time derivatives are extremely small. An alternative is therefore to observe the effect of gravitational waves right where they are generated.

In 1974 the American astronomers R. Hulse and J. Taylor discovered are radio pulsar, PSR1913-16, which is part of a binary system of apparently compact stars (neutron stars). Taylor and others observed PSR1913-16 over several years. They were able to measure the steady shortening of the period of the binary system due to the above energy loss or gravitational radiation damping (see the box below). The comparison of their measurement with the theory was clear. Hulse and Taylor received the Nobel Prize for their discovery and the subsequent verification of Einstein's prediction of gravitational waves in 1993.

● **Gravitational Radiation:** In electrodynamics we had calculate the total energy emitted from a source (antenna) in the dipole approximation. The same can be done here. However, because a mass distribution cannot have a dipole moment, the leading term is a quadrupole term. The result for the emitted power generated by two masses orbiting each other is

$$-\frac{dE}{dt} = \frac{32 \, G (\mu r^2)^2 \Omega^6}{5c^5} . \tag{5.75}$$

Here μ is the reduced mass, r is the orbital radius, i.e. μr^2 is the moment of inertia, and Ω the rotation frequency. Note that we have not used $c = 1$ here. Aside from the factor $32/5$, which requires a detailed calculation, cf. Sect. 110

---

[2]There is no phase shift here because we neglect friction.

in [12], we can obtain this formula from simple dimensional analysis. For the system consisting of the Earth and the Sun we find -dE/d$t \approx 200$ W. The tiny amount indicates that much more massive bodies are needed—like PSR1913-16.

We do not want to present a detailed calculation of emission of gravitational waves and refer the reader to the sources in the preface. Instead a few rough considerations must suffice. If we choose (4.48) as our starting point, then we can calculate the emission of waves from a source analogous to the emission of electromagnetic radiation (cf. [9]), i.e.

$$\bar{h}_{\tau\sigma} = \frac{4G}{R} \int (\mathcal{T}_{\tau\sigma})_{t-R} dV . \tag{5.76}$$

Here R is the distance between observer and emitter. The subscript $t - R$ is a mere reminder of the retardation, i.e. we detect radiation emitted at an earlier time. Obviously the integral has the dimension of energy. Let's for instance assume there are two typical masses m, separated by the distance r, revolving around each other with frequency Ω. The product $mr^2\Omega^2$ has units of energy. So all together we expect (without using $c = 1$)

$$h \sim \frac{G}{Rc^4} mr^2\Omega^2 . \tag{5.77}$$

It is easy to check that the expression is dimensionless as it should be.

Let us estimate the expected magnitude of h in the case of the PSR1913-16 binary system. Its distance from Earth is about 21000 light-years or $R \approx 2 \cdot 10^{20}$ m and the masses involved are roughly 1.4 solar masses, i.e. $m \approx 3 \cdot 10^{30}$ kg. The period of rotation of the binary is 7.75 hours or $\Omega \approx 2.2 \cdot 10^{-4}$ s^{-1}. Calculating r according to $Gm/r^2 = m\Omega^2 r$ we find $r \approx 2 \cdot 10^{9}$ m. Inserting these numbers into (5.76) yields $h \sim 10^{-23}$. It is not really surprising that Weber didn't detect this. In addition, the resonance frequency of his aluminium detectors is in the kHz range and therefore very much different from the above Ω.

The state of the art operational gravitational wave detectors at the time of writing are the two detectors of the Advanced Laser Interferometer Gravitational-Wave Observatory (Advanced LIGO) and the Virgo detector. Each of the Advanced LIGO detectors features two interferometer arms that have a length of $\epsilon = 4$ km and a sensitivity to amplitudes of only $\approx 10^{-19}$ m, which is four orders of magnitude smaller than an atomic nucleus. This sensitivity is achieved by repeated reflection of a laser beam in the interferometer arms, boosting the distance traveled by the beams to 1200 km and significantly increasing the laser energy. The mirrors themselves serve as test masses and are suspended in vacuum by a sophisticated multi-stage damping mechanisms to shield them from environmental noises such as earthquakes or trucks going by. When in operation, electrostatic actuators exert tiny forces on the mirrors, displacing them slightly so as to always keep the interferometer at a point of destructive interference. When the background effects are subtracted, the force required to maintain destructive interference is therefore proportional to the gravitational wave strain and thus constitutes the signal. The three interferometers of the Advanced LIGO and Virgo detectors as well as planned upcoming additions are distributed around the globe so as to triangulate the sources of the gravitational waves via time of arrival differences. They also have the interferometer arms pointing in different directions in space to infer the polarization of the incoming gravitational waves.

Let us now see what is the sensitivity of these detectors terms of h. We use (5.61), i.e.

$$\frac{1}{2}h_{\text{LIGO}}\epsilon = 10^{-19}\,\text{m} \tag{5.78}$$

to find for the amplitude $h_{\text{LIGO}} \approx 10^{-22}$, which is very small indeed. In principle this is close to what is required to detect gravitational waves from PSR1913-16. However, here the real problem is the rather long period. LIGO's optimal sensitivity is in a frequency range around 100 Hz. A gravitational wave with this frequency has a wavelength of about $\lambda_{gw} = c/100\,\text{Hz} \approx 3000\,\text{km}$.

One can wonder whether gravitational waves of such high frequency do exist. After all a frequency of 100 Hz implies two heavy masses encircling each other in a matter of milliseconds. There is however a class of astronomical objects, namely merging binary black holes and neutron stars, that produce these kinds of signals and that are sufficiently common and powerful so that we can detect them. A growing number of events of this sort are observed, and on human scales their proportions are startling. In the final stages of a typical binary black hole merger, two black holes of ∼30 solar masses orbit each other in a few tens of milliseconds, radiating off an energy equivalent to a few solar masses within fractions of a second (see Problem 3).

At this point somebody may wonder about the effect of the gravitational wave on the laser light in the interferometer. Is it not red- or blueshifed, too? The situation is similar to the freely falling observer in the Pound–Rebka experiment discussed in Sect. 2.3.2. For 'his' light there is no effect. More precisely the metric in his frame of reference is $g_{\alpha\beta} = \eta_{\alpha\beta} + \mathcal{O}(h(\epsilon/\lambda_{gw})^2)$ (**). Thus, as long as $(\epsilon/\lambda_{gw})^2 \ll 1$, we should expect no significant effect on the wavelength of the light as perceived by the observer at the corner. In the LIGO-case and with the above numbers $(\epsilon/\lambda_{gw})^2 \approx 10^{-6}$![3]

(**): How does this correction arise? Consider the following simple one-dimensional example for deviation from flatness. Figure 5.5 shows a circle of radius R touching the x-axis at the origin. The x-axis is a 'tangent space' to the circle near the origin. The equation for the bottom half of the circle is $y_-/R = 1 - \sqrt{1 - (x/R)^2}$ and near the origin we have $\delta y_- \approx \frac{1}{2}\delta x^2/R$. A small piece of arc length is given by $\delta s \approx \sqrt{\delta x^2 + \delta y_-^2}$ or $\delta s \approx \delta x(1 + \frac{1}{8}(\delta x/R)^2)$. Returning to space time we have

---

[3]Note that here ϵ is indeed 4 km and not 1200 km.

Fig. 5.5 Deviation from
flatness

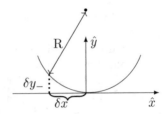

$1/R^2 \sim |\mathcal{R}_{\alpha\beta\gamma\delta}| \sim \ddot{h}^{TT} \sim \Omega^2 h^{TT}$ (cf. (4.42) and (5.73)). Using $\Omega \sim 1/\lambda_{gw}$ completes the argument.[4]

● **Example - Gravitational Wave Detection:** On August 14, 2017 LIGO
(Laser Interferometer Gravitational-Wave Observatory) detected gravitational
radiation emitted during the final moments of the merger of two black holes.
The figure below shows the oscillations seen by LIGO's Hanford detector.
The estimated masses of the two black holes were about 31 and 25 solar
masses, respectively. The merger occurred about 1.8 billion light-years away.
Use the above order-of-magnitude-method to estimate the attendant maximum
distortion amplitude of the 4000 m mirror spacing during the event. Assume
that the gravitational wave travels perpendicular to the detector plane.

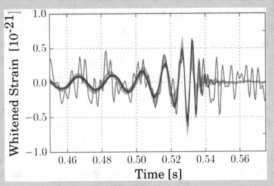

Image courtesy of the LIGO Scientific Collaboration and the Virgo Collaboration.
Shortly before their merger the separation of the black holes, r, can be
estimated via their Schwarzschild radius, i.e.

$$\frac{2GM}{c^2 r} \sim 1 . \tag{5.79}$$

---

[4]A discussion of this and numerous other aspect of gravitational waves, their generation and
detection can be found here—https://cosmolearning.org/courses/overview-of-gravitational-wave-
science-400/.

Here we include c explicitly. With $M \approx 30 M_\odot$ this yields $r \sim 9 \cdot 10^4$ m. The frequency at this point (around 0.53 s in the graph) is about 150 Hz, i.e. $\Omega \approx \frac{1}{2} 2\pi\, 150$ Hz (*). Using the relation (5.77) and $R \sim 1.8 \cdot 10^9$ lt yr or $\approx 1.8 \cdot 10^{25}$ m yields

$$h_{max} \sim 0.1 \cdot 10^{-21} . \tag{5.80}$$

The attendant maximum distortion of the mirror separation is

$$4000 \text{ m } h_{max} \sim 0.4 \cdot 10^{-18} \text{ m} . \tag{5.81}$$

Of course, all this is very rough.

(*) The factor $\frac{1}{2}$ requires an explanation. Consider two bodies orbiting each other according to Newton's law. Their orbit can be described in terms of a radial coordinate r and a polar coordinate ψ. Here $\dot{\psi} = \Omega$. The radiation detected by LIGO is quadrupole radiation. This radiation is governed by mass quadrupole moments, containing products like $\cos^2 \psi$ or $\sin \psi \cos \psi$, with a doubled period compared to $\cos \psi$ or $\sin \psi$. It is also easy to intuitively understand this difference to an electromagnetic dipole radiation, because for gravity both charges are positive.

5.6 Problems

1. The central black hole of our milky way, Sagittarius A*, has a mass of approximately four million solar masses. The star S14 orbiting it has a semimajor axis of $a \approx 1680$ AU and an excentricity of $e \approx 0.974$. According to our approximation, what is its periastron precession per revolution and per century?
2. RX J1131-1231 is a celestial object that consists of a quasar and a galaxy which line up along our line of sight. The object closer to us, the galaxy at a distance of $d_0 \approx 3.5 \times 10^9$ light-years, is deflecting the light from the quasar at a distance of $d_1 \approx 6.3 \times 10^9$ light-years, so that it looks to us like a ring with an angular diameter of $\theta \approx 3''$. This phenomenon is called an Einstein ring.

   (a) Estimate the mass of the foreground galaxy.
   (b) Now assume that the objects do not lie directly behind each other, but that the angle between their centers is $\varphi \approx 0.4''$. At some time you observe one side of the ring suddenly becoming brighter, because the quasar becomes brighter. How much time will it take before the other side becomes brighter, too?

   Hint: For this exercise ignore cosmic expansion.

3. Consider the metric

$$d\tau^2 = dt^2 - (1 + 2h_+(t - z))dx^2 - 4h_\times(t - z)dxdy$$
$$-(1 - 2h_+(t - z))dy^2 - dz^2 ,$$

describing a gravitational wave in z-direction to first order in the small perturbations $h_+ \ll 1$ and $h_\times \ll 1$. Both h_+ and $h_\times$ are functions of $t - z$, i.e. waves
propagating in the positive z direction.
Hint: Note that $h_+ = h_{xx}^{TT}$ and $h_\times = h_{xy}^{TT}$.

(a) Compute, up to second order in h_+, $h_\times$ the inverse metric and the Christoffel
    symbols.
(b) Show that up to first order in h_+, $h_\times$ the vacuum Einstein equation is satisfied.
(c) Now repeat the calculation and compute $G_{\mu\nu}$ up to second order in h_+, $h_\times$.
    The result will not be zero anymore, because this goes beyond the linear
    approximation. Compute the $T_{\mu\nu}$ that would produce such a $G_{\mu\nu}$. One can
    interpret this $T_{\mu\nu}$ (suitably averaged over many oscillation periods) as the
    energy momentum tensor carried by the gravitational wave.
    Hint: the final result for the energy density is $T^{00} = \frac{1}{4\pi G}(2h_+ h_{+,tt} + h_{+,t}^2 + 2h_\times h_{\times,tt} + h_{\times,t}^2)$
(d) Let us apply these results to the famous gravitational wave signal GW150914
    that resulted from a merger of two black holes.

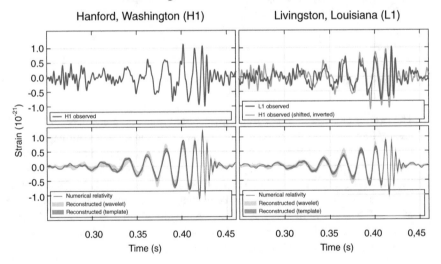

Image courtesy of the LIGO Scientific Collaboration and Virgo Collaboration.

The plot represents the gravitational strain (essentially h) versus the time
recorded with two detectors. The event occurred at a distance from Earth of
about 410 Mpc. Use the results for T^{00} you obtained in the previous exercise
to give an order of magnitude estimate of the total energy that was radiated
off during the last ~ 0.2 s before the merger.

4. Show that when one rotates the coordinate system about the gravitational waves'
   propagation direction (the z-direction) by an angle ϕ (so that $x' + iy' = (x + iy)e^{-i\phi}$), the gravitational wave fields h_+ and $h_\times$ transform according to the
   equation

$$h'_+ + ih'_\times = (h_+ + ih_\times)e^{-i2\phi} .$$

This equation is often described by saying that $h_+ + ih_\times$ has spin-weight 2, i.e.
the graviton has spin 2.

Chapter 6
Black Holes

For strong gravitational fields, general relativity predicts phenomena which are entirely alien to Newtonian gravity, most notably black holes. This chapter explores their properties and formation for the simplest case, a non-spinning and uncharged Schwarzschild black hole.

6.1 A Closer Look at the Schwarzschild Solution

Probably you have noticed in the last chapter that the Schwarzschild solution (5.13) becomes singular at $r = 2GM$, where the quantity $r_s = 2GM$ is called the Schwarzschild radius. The singularity is irrelevant in the case of the Sun since its Schwarzschild radius is a mere 2.95 km and the Schwarzschild solution is only valid outside the spherical mass distribution. Even for such extreme objects as white dwarfs and neutron stars the Schwarzschild radius corresponding to their total mass is still smaller than their actual radius.

Objects that do not extend beyond their Schwarzschild radius are called black holes. There is now overwhelming evidence that black holes do exist, some of them extremely massive and in the centers of galaxies. For the simple case, when these black holes do not interact with their surroundings, do not rotate and are electrically neutral, they are described by the Schwarzschild metric. This idealised case is known as the 'eternal Schwarzschild black hole', which we now study in some detail.

Per construction the Schwarzschild metric

$$d\tau^2 = \left(1 - \frac{r_s}{r}\right) dt^2 - \frac{dr^2}{1 - \frac{r_s}{r}} - r^2(d\theta^2 + \sin^2\theta d\phi^2) \tag{6.1}$$

goes over into the flat metric of free space at large distances $r \gg r_s$. Far from the center, t is the time as we know it and r, θ and ϕ are simply the usual spherical coordinates. Furthermore, the angular part of the metric does not deviate from spherical

© Springer Nature Switzerland AG 2020
R. Hentschke and C. Hölbling, *A Short Course in General Relativity and Cosmology*, Undergraduate Lecture Notes in Physics,
https://doi.org/10.1007/978-3-030-46384-7_6

coordinates even as we go to smaller r. The surface area of a sphere at coordinate r is always $4\pi r^2$ as it is in flat space. So the interesting questions are, how do radial distances and time change when we approach $r = r_s$ or when we cross it?

To answer these questions, let us first imagine two events that happen at the same r, θ and ϕ coordinates and are separated in their t coordinate by dt. The metric (6.1) tells us that the proper time between these two events is

$$d\tau^2 = \left(1 - \frac{r_s}{r}\right) dt^2 , \tag{6.2}$$

i.e. the time between the two events is dt at large distances $r \gg r_s$ and shrinks as we approach r_s. If we go to $r = r_s$ however, things start getting weird. At $r = r_s$, the proper time $d\tau = 0$ so the events are no more separated in a timelike manner, but lightlike. For $r < r_s$ we even get $d\tau^2 < 0$, which tells us that the events are spacelike separated. This means the two events happen at different places and there is no way to go from one event to the other. Or, in other words, you cannot remain at a constant 'position' (r, θ, ϕ) if you are inside the Schwarzschild radius.

Since t has become a spatial coordinate inside the Schwarzschild radius, there is the obvious question about what the time coordinate is. To answer this, let us next look at two events that happen at the same coordinates t, θ and ϕ but that are separated in the coordinate r by dr. From the metric (6.1) we see that the proper distance between these two events is

$$-d\tau^2 = \frac{dr^2}{1 - \frac{r_s}{r}} . \tag{6.3}$$

Far from the Schwarzschild radius, i.e. $r \gg r_s$, there is a distance dr between the two events. This distance increases and actually diverges at $r = r_s$, but inside we see that $d\tau^2 > 0$ and the events are timelike separated. So for $r < r_s$, the coordinate r represents time and the coordinate t represents a spatial distance. The coordinates r and t have swapped places. Inside the Schwarzschild radius, the position of an object is given by t, θ, ϕ and the time is given by r. There is one catch however: the coordinate r starts (or ends) at $r = 0$. We now have two choices: either time runs forward in r, which implies that objects can emerge from one point (remember that the surface area of a sphere at coordinate r is always $4\pi r^2$), or time runs forward in $-r$ and all objects will end in one point. In either case, we can actually compute how long it takes from $r = r_s$ to $r = 0$ or the other way around if one stays at fixed t, θ, ϕ. The proper time from the Schwarzschild radius to the center is

$$\tau = \int_{r_s}^0 \frac{-dr}{\sqrt{\frac{r_s}{r} - 1}} \stackrel{r/r_s = \sin^2 y}{=} 2r_s \int_0^{\pi/2} \sin^2 y \, dy = r_s \frac{\pi}{2} , \tag{6.4}$$

which is finite. Thus, if we are falling in, cross the Schwarzschild radius and remain at fixed (t, θ, ϕ), we will end up squished to a point in a time $r_s \frac{\pi}{2}$. But what if we move? Since all other coordinates are spacelike, they will only decrease $d\tau$ and the proper time will be shorter. Thus, as a general statement, having crossed the Schwarzschild radius, we at most have a time $r_s \frac{\pi}{2}$ left before being squished to a point.

There is of course nothing special about anything that falls in beyond the Schwarzschild radius, including light rays. So if we try sending a radio signal or another kind of signal out from inside r_s, it will end up at $r = 0$ all the same. In this sense $r = 0$ is a singularity. It is not a point in space, but rather the end of the time evolution of everything that at some point crossed the Schwarzschild radius. There is a horizon located at $r = r_s$ and nothing from inside the horizon can reach the world outside anymore, hence the term black hole.

Since these results seem strange at first sight, you might suspect (as a lot of physicists did for many decades) that black holes are unphysical. There are two obvious points of critique: The singularity in the metric at $r = r_s$, that we have completely glossed over, and the question whether black holes can actually form. We will look at both points in some detail and see that the Einstein equations can describe the formation of black holes without encountering singularities.

But what about the possibility we mentioned above that time does not end at $r = 0$ but instead starts there? The resulting objects are called white holes and they are just the time reverse of black holes. They too have a horizon but it forbids any object entering it and everything inside the horizon is expelled in a proper time no longer than $r_s \frac{\pi}{2}$. Therefore white holes are believed to be unstable and no mechanism of producing them is known that is not highly speculative.

6.2 The Schwarzschild Black Hole in Various Coordinates

The key to resolving the singularity at $r = r_s$ in the Schwarzschild metric (6.1) is the realization that the coordinates t and r are inherited from the distant observer. If you are falling past a horizon, your clocks and rods are so out of sync with the asymptotic observer that you cannot even agree on what is a clock and what is a rod anymore. We will therefore rewrite the Schwarzschild metric in various different coordinates that are more suited for observers falling through the horizon or hovering slightly above it. Before doing this however, we will look at flat space and see that, given proper coordinates, we can produce horizons even there.

6.2.1 Rindler Space

The top panel in Fig. 6.1 shows trajectories (green lines) of uniformly accelerated observers in Newtonian physics. They move on curves

$$x = \frac{1}{2}at^2 + b_o , \tag{6.5}$$

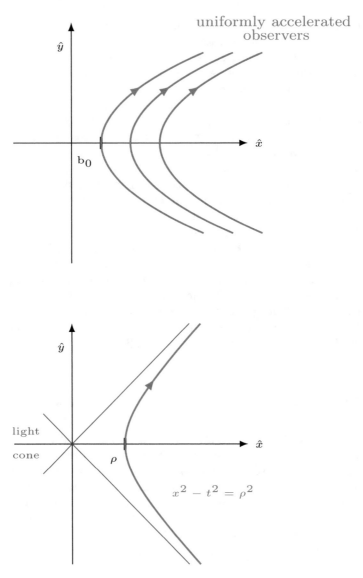

Fig. 6.1 Top: uniformly accelerated observers in Newtonian physics; bottom: uniformly accelerated observer in local Lorentz frame

Fig. 6.2 The (sad) story of
Alice and Bob

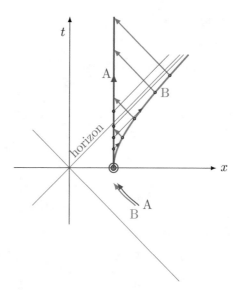

where a is the uniform or constant acceleration. The quantity b_o is the value of x when $t = 0$. The bottom panel in Fig. 6.1 shows a trajectory described via

$$x^2 - t^2 = \rho^2 . \qquad (6.6)$$

Again ρ is the value of x when $t = 0$. Taking the square of (6.5) and comparing the result with (6.6) in the limit of small t, i.e. the observer in top panel in Fig. 6.1 moves slowly, we find

$$\rho = \frac{1}{a} . \qquad (6.7)$$

We also note that $x^2 - t^2$ is invariant under the Lorentz transformations (2.10) ($c = 1$!), i.e.

$$x^2 - t^2 = x'^2 - t'^2 . \qquad (6.8)$$

This tells us that the green hyperbola in the bottom panel of Fig. 6.1 consists of points each corresponding to local Lorentz frames which each accelerate with the same constant acceleration. Thus, all in all the green line in this panel describes the trajectory of a uniformly accelerating observer in relativistic flat spacetime.

Figure 6.2 tells the story of two observers, Alice and Bob. Alice and Bob approach together in their lab or elevator cabin. The lab slows down and at $t = 0$ Alice gets off while Bob stays in the lab. The lab picks up speed and Bob is moving away from Alice, who is at rest in our coordinate system, with constant acceleration.

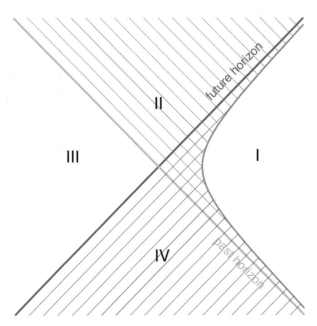

Fig. 6.3 Four regions of flat spacetime for an accelerated observer

Alice and Bob communicate continuously by exchange of electromagnetic messages. In the figure these signals are the arrows going back and fourth between the
(vertical) trajectory of Alice and the (hyperboloid) trajectory of Bob. However there
is a limit to this exchange. Bob will not be able to receive any communication from
Alice once she crosses the red line—the horizon. He will continue to send messages
which will of course reach Alice's trajectory in her unaccelerated reference frame,
but nothing will ever come back from her. Note that Bob's separation from the origin
remains spacelike throughout and thus no point on the horizon ($x = t$) can ever be
in his past. So, from his perspective he will never see her crossing the horizon. From
her perspective crossing the horizon is just another ordinary moment in time without
any particular significance—or danger. Thus, here we have a simple model, which
nevertheless incorporates an event horizon.

In fact, there is even more than one horizon here. In Fig. 6.3 Bob's worldline is
shown together with signals he can receive and send. We can see that nothing from
behind the future horizon can reach him, but there is also the past horizon, drawn
in orange, beyond which he cannot send any signal (assuming of course that he has
always been accelerating). These two horizons divide up flat spacetime into four
regions for a constantly accelerated observer: region I, where he can send signals
to and receive signals from, region II where he can send signals to but not receive
signals from, region III where he can neither send nor receive signals and region IV
where he can receive signals from but not send signals to.

We now exchange t and x for the coordinates introduced previously in (2.28). Computing the proper time we obtain

$$d\tau^2 = dt^2 - dx^2 = \rho^2 d\alpha^2 - d\rho^2 . \tag{6.9}$$

The coordinates ρ and α are known as Rindler[1] coordinates and they are appropriate coordinates for observers like Bob with a constant acceleration $1/\rho$ and a fixed distance ρ from the horizon. For $\rho > 0$ they describe the region I of Fig. 6.3 which is also known as the Rindler wedge.

There is one more useful set of coordinates which we want to introduce in flat space first: light cone coordinates. The idea behind them is to dispose of spacelike and timelike coordinates which might change meaning at a horizon and replace them by coordinates that are lightlike everywhere. We introduce the coordinates

$$v = t + x \qquad u = t - x \tag{6.10}$$

and thus the metric becomes

$$d\tau^2 = dudv - dy^2 - dz^2 . \tag{6.11}$$

We can also express the light cone coordinates in terms of Rindler coordinates. They are

$$v = \rho e^{\alpha} \qquad u = -\rho e^{-\alpha} . \tag{6.12}$$

Very conveniently, the axes of light cone coordinates $u = 0$ and $v = 0$ represent the past and future horizon of a constantly accelerated Rindler observer as depicted in Fig. 6.4. Note also that every timelike or lightlike trajectory that starts inside the future light cone $v > 0, u > 0$ (the region II of Fig. 6.3) will stay there. This statement is trivial in flat space, but it also holds for black holes, where it is very useful.

6.2.2 Near Horizon Coordinates

Let us now return to the Schwarzschild solution (6.1) and check if we can learn something about its horizon or singularity in different coordinates. As a first step, we place ourselves closely above the horizon and replace the radial coordinate r by the proper distance to the horizon ρ. This allows us to write the metric as

$$d\tau^2 = \left(1 - \frac{r_s}{r(\rho)}\right) dt^2 - d\rho^2 - r(\rho)^2 d\Omega^2 \tag{6.13}$$

To obtain $r(\rho)$, we compute the horizon distance ρ in Schwarzschild coordinates

---

[1] Rindler, Wolfgang, American physicist, *Vienna 18.5.1924, †Dallas 8.2.2019.

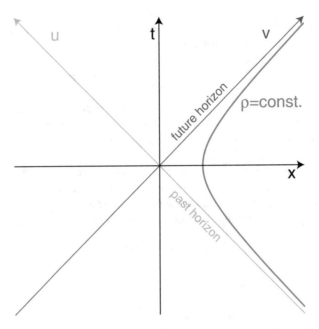

Fig. 6.4 Light cone coordinates in flat spacetime. The trajectory of a constantly accelerated observer (with constant proper distance ρ to its Rindler horizon) is also depicted

$$\rho = \int_{r_s}^{r} \frac{dr'}{\sqrt{1 - \frac{r_s}{r'}}} = \sqrt{r(r - r_s)} + r_s \operatorname{atanh}\sqrt{1 - \frac{r_s}{r}} \tag{6.14}$$

Near the horizon $(1 - r_s/r) \ll 1$ and so

$$\rho \simeq 2\sqrt{r_s(r - r_s)} \qquad r \simeq r_s + \frac{\rho^2}{4r_s} \,, \tag{6.15}$$

which allows us to write the metric near the horizon as

$$d\tau^2 \simeq \rho^2 \left(\frac{dt}{2r_s}\right)^2 - d\rho^2 - r(\rho)^2 d\Omega^2 \,. \tag{6.16}$$

By a simple rescaling of the time coordinate $\alpha = t/2r_s$ we see that the radial part is just the Rindler metric (6.9). So an observer hovering above the horizon at a fixed distance ρ is locally identical to the constantly accelerating Bob in our flat space example. The acceleration is now caused by the black hole, which is just an example of the equivalence principle. As a final step, we can introduce local Cartesian coordinates

$$X = \rho \cosh \alpha \quad \text{and} \quad T = \rho \sinh \alpha \,. \tag{6.17}$$

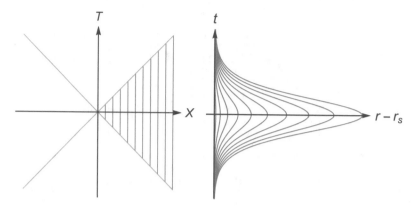

Fig. 6.5 Inertial (i.e. freely falling) observers near the horizon of a black hole. The left panel shows the worldlines of a family of radially freely falling observers in near horizon coordinates, while the right one shows the same worldlines in Schwarzschild coordinates (using $r - r_s \approx \rho^2/(4r_s)$ and $t = 2r_s\alpha$ in conjunction with (6.17)). The worldlines are characterized by a constant near horizon coordinate $X = X_i$ where the X_i are equally spaced. Notice how inertial observers move along a straight worldline in locally free falling near horizon coordinates while their trajectories are curved in Schwarzschild coordinates, which are not inertial. On the left panel, the orange lines show the position of the horizon, while on the right panel the horizon is at the t-axis

With these the metric is locally flat

$$d\tau^2 \simeq dT^2 - dX^2 - r(\rho)^2 d\Omega^2 . \tag{6.18}$$

We get rid of the gravitational pull of the black hole locally by letting our coordinate system freely fall into it, which again is just the equivalence principle. This is Alice's perspective from the flat space example and again nothing special happens from this perspective when crossing the horizon.

It is a simple but instructive exercise to draw the world lines of various observers in near horizon coordinates and compare them to the same world lines in Schwarzschild coordinates. In Fig. 6.5 we have plotted world lines of freely falling observers at constant X in both coordinates, while in Fig. 6.6 we show world lines of observers hovering at a fixed distance above the horizon. It is evident that the descriptions differ dramatically in the different coordinate systems.

6.2.3 Eddington–Finkelstein Coordinates

We have now seen that locally the horizon of a Schwarzschild black hole looks like the Rindler horizon of an accelerated observer and nothing special happens from the perspective of a free falling observer crossing it. However, since we also saw that from the outside it looks like nothing crosses the horizon, the question whether

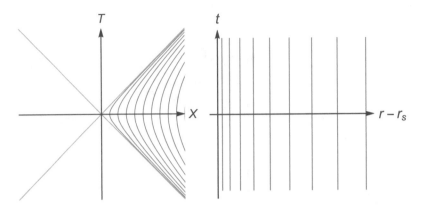

Fig. 6.6 Observers hovering at a constant distance above the horizon. The left panel shows a family of worldlines of these observers in near horizon coordinates, while the right one shows the same worldlines in Schwarzschild coordinates. The worldlines are characterized by a constant proper distance from the horizon $\rho = \rho_i$ where the ρ_i are equally spaced. The curved world lines in free falling horizon coordinates explicitly show that these observers are constantly accelerating. On the left panel, the orange lines show the position of the horizon, while on the right panel the horizon is at the t-axis

a black hole can form at all seems even more pressing. In order to approach this question, we introduce yet another coordinate system which is specifically suited for radially infalling light rays.

We start from the Schwarzschild solution (6.1) and introduce the coordinate

$$r^* = r + r_s \ln\left(\frac{r}{r_s} - 1\right) \tag{6.19}$$

that pushes the horizon to negative infinity $r^* \xrightarrow{r \to r_s} -\infty$. It has the property that

$$\frac{1}{1 - \frac{r_s}{r}} dr^2 = \left(1 - \frac{r_s}{r}\right) dr^{*2} , \tag{6.20}$$

which allows us to rewrite the Schwarzschild metric as

$$d\tau^2 = \left(1 - \frac{r_s}{r}\right)(dt^2 - dr^{*2}) - r^2(d\theta^2 + \sin^2\theta d\phi^2) . \tag{6.21}$$

The advantage of writing the metric in this form becomes evident when we try to understand a light ray radially falling into or emerging from a black hole. For light rays we have $d\tau = 0$, which for constant angular coordinates implies $dt = \pm dr^*$.

Let us now concentrate on a light ray that is falling radially into a black hole. For this light ray we have $dt = -dr^*$, so if we define a new coordinate $v = t + r^*$, this coordinate is constant along the light ray. All radially ingoing light rays thus have constant coordinates v, θ and ϕ and a point on the light ray can be given either by r^*

or t. We choose r^* and eliminate t. Since $dt = dv - dr^*$, we can write the metric as

$$d\tau^2 = \left(1 - \frac{r_s}{r}\right)(dv^2 - 2dr^* dv) - r^2(d\theta^2 + \sin^2\theta d\phi^2). \qquad (6.22)$$

Using (6.20) we can now replace r^* by r again and cast the metric in the form

$$d\tau^2 = \left(1 - \frac{r_s}{r}\right)dv^2 - 2drdv - r^2(d\theta^2 + \sin^2\theta d\phi^2), \qquad (6.23)$$

which is known as the (ingoing) Eddington[2]–Finkelstein[3] metric. Note that although we derived this metric from the Schwarzschild metric, it contains no singularity at $r = r_s$. The coordinate r just describes where a point is located on a radially ingoing light ray (v, θ, ϕ) and its transformation from being spacelike to being timelike at $r = r_s$ does not pose any particular problem.

6.2.4 Kruskal–Szekeres Coordinates

We will soon see that Eddington–Finkelstein coordinates are very well suited to describe the formation of a black hole. To see the structure of spacetime with a Schwarzschild black hole however we can do even better. The coordinate $v = t + r^*$ in (6.21) is a radial, outward pointing light cone coordinate. This is very convenient when we draw trajectories of objects in such coordinates, since nothing outgoing can ever overtake the outgoing light ray. We can do the same for infalling light rays by replacing r^* with the second light cone coordinate $u = t - r^*$ so that

$$dudv = dt^2 - dr^{*2}. \qquad (6.24)$$

If we follow the trajectory of any object or light ray, neither u nor v can ever decrease. We can thus easily read off from a diagram in such coordinates all the possibilities where an object or light ray may go.

Writing the metric (6.21) in the new coordinates yields

$$d\tau^2 = \left(1 - \frac{r_s}{r}\right)dudv - r^2(d\theta^2 + \sin^2\theta d\phi^2). \qquad (6.25)$$

Remember that the horizon was removed to $r^* \to -\infty$, so in the coordinates u and v it is also not at any finite value. If we want to bring the horizon back to finite values of the coordinates, we can simply exponentiate u and v and define

$$U = -e^{-\frac{u}{2r_s}} \quad \text{and} \quad V = e^{\frac{v}{2r_s}}. \qquad (6.26)$$

---

[2]Eddington, Arthur Stanley, British astronomer, physicist and mathematician, *28.12.1882 Kendal, †22.11.1944 Cambridge.

[3]Finkelstein, David, American physicist, *New York City 19.7.1929, †Atlanta 24.1.2016.

We note in passing the curious mathematical fact that with this definition, both U and V are unchanged if one replaces t by $t + i4\pi r_s$. Since we have

$$dU = \frac{1}{2r_s}e^{-\frac{u}{2r_s}}du = -\frac{1}{2r_s}Udu \quad \text{and} \quad dV = \frac{1}{2r_s}e^{\frac{v}{2r_s}}dv = \frac{1}{2r_s}Vdv \; , \quad (6.27)$$

we can write the metric (6.25) as

$$d\tau^2 = -\frac{4r_s^2}{UV}\left(1 - \frac{r_s}{r}\right)dUdV - r^2(d\theta^2 + \sin^2\theta d\phi^2) \; . \quad (6.28)$$

This is the Schwarzschild metric in yet another set of coordinates that are known as the light cone Kruskal[4]–Szekeres[5] coordinates. The nice thing about them is that they inherited the property of being light cone coordinates from u and v. It means that curves of constant V represent radially infalling light rays while curves of constant U, at least for $r > r_s$, represent radially outgoing light rays. Similar to the flat space case, all timelike or lightlike trajectories from a point lie inside its future light cone which is just given by lines of constant U and constant V (usually drawn diagonally) emanating from it as depicted in Fig. 6.7. Thus going into the future in Kruskal–Szekeres coordinates means going upwards inside this light cone.

Let us now see where the horizon lies in these coordinates. For the product UV we find

$$UV = -e^{\frac{v-u}{2r_s}} = -e^{\frac{r^*}{r_s}} \quad (6.29)$$

which, using (6.19), simplifies to

$$UV = -\left(\frac{r}{r_s} - 1\right)e^{\frac{r}{r_s}} \quad (6.30)$$

so the horizon $r = r_s$ lies at $UV = 0$. This allows us to write the metric (6.28) in the form

$$d\tau^2 = \frac{4r_s^3}{r}e^{-\frac{r}{r_s}}dUdV - r^2(d\theta^2 + \sin^2\theta d\phi^2) \; , \quad (6.31)$$

which makes explicitly clear that there is no singularity except at $r = 0$. It is worth pointing out here that r/r_s is a function of the product UV only, in fact it is the solution of the transcendental equation (6.30), which one can write as $r = r_s(1 + W_0(-UV/e))$ where W_0 is the principle branch of the Lambert W function. The final form of the metric in terms of Kruskal–Szekeres coordinates U and V therefore is

---

[4]Kruskal, Martin David, American mathematician and physicist, *28.9.1925 New York City, †26.12.2006 Princeton.

[5]Sekeres, George, Hungarian-Australian mathematician, *29.5.1911 Budapest, †28.8.2005 Adelaide.

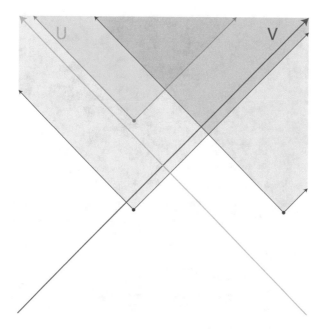

Fig. 6.7 Future light cones of various events in light cone Kruskal–Szekeres coordinates

$$d\tau^2 = r_s^2 \left(\frac{4e^{-1-W_0\left(-\frac{UV}{e}\right)}}{1 + W_0\left(-\frac{UV}{e}\right)} dU \, dV \right.$$
$$\left. - \left(1 + W_0\left(-\frac{UV}{e}\right)\right) \left(d\theta^2 + \sin^2\theta d\phi^2\right) \right). \tag{6.32}$$

In close analogy to flat space for an accelerated observer, we have a past horizon at $V = 0$ and a future horizon at $U = 0$ which split the spacetime into four distinct regions as displayed in Fig. 6.8. Let us try to understand these regions one by one, starting with region I. In region I we have $U < 0$ and $V > 0$ so $UV < 0$ and thus, according to (6.30), $r > r_s$. This is the region outside the black hole and trajectories of observers at a constant distance above the horizon look like the trajectories of constantly accelerated observers in flat space. We have drawn three such trajectories in Fig. 6.8 corresponding to $r = \{1.2, 1.4, 1.6\}r_s$. Starting from this region objects and ingoing light rays can cross the future horizon into region II, but not into regions III or IV. Conversely, the only other region from which region I can be reached is region IV. Objects and light rays from there can cross over the past horizon into I.

Next we look at region II. Here $U > 0$ and $V > 0$, so $UV > 0$ and thus, according to (6.30), $r < r_s$ and we are inside the black hole. Now we can see the usefulness of Kruskal–Szekeres coordinates. Since radial light rays are diagonal and no trajectory can ever leave its future light cone, we can read off directly from Fig. 6.8 that region II can be reached from all other regions but once something is inside region II

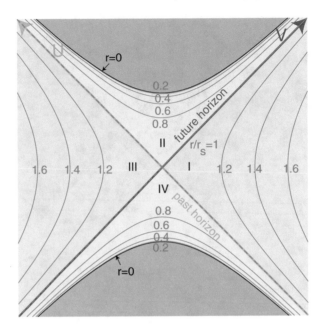

Fig. 6.8 A Schwarzschild black hole in light cone Kruskal–Szekeres coordinates. Lines of constant coordinate r are highlighted

it remains there. We can further see that $r = 0$ according to (6.30) corresponds to the hyperbola $UV = 1$. There spacetime ends in a singularity and the grey region above this hyperbola in Fig. 6.8 is not part of the Schwarzschild spacetime. Further curves at constant $r = \{0.2, 0.4, 0.6, 0.8\}r_s$ are also depicted. Drawn in Kruskal–Szekeres coordinates it is obvious that they are spacelike and thus cannot represent the trajectory of any object. It is furthermore very transparent in Kruskal–Szekeres coordinates that everything that has crossed over into region II will proceed to smaller r and ultimately end up in the singularity. We already know that the singularity at $r = 0$ is a single point physically although it is a hyperbola in Fig. 6.8. This stretching of a single point is due to our choice of coordinates and generically it is the price we pay for the convenience of radial light rays being diagonal in the diagram.

With regions I and II we have thus already covered the entire black hole. Region II is the inside of the black hole and region I is outside, so what could regions III and IV possibly be? To get a first idea let us again turn to (6.30). In region III we have $U > 0$ and $V < 0$, i.e. $UV < 0$, and therefore this region is outside the Schwarzschild radius $r > r_s$. Similarly, in region IV we have $U < 0$ and $V < 0$, so $UV > 0$ and thus $r < r_s$, so region IV is inside the Schwarzschild radius. Mathematically, we can obtain regions III and IV by simply reversing the signs of both U and V in regions I and II. Additionally, the metric (6.31) only depends on the products UV and $dU dV$, as we noted above, and not on the individual signs of U and V. Does this mean we just have another copy of the black hole interior and exterior?

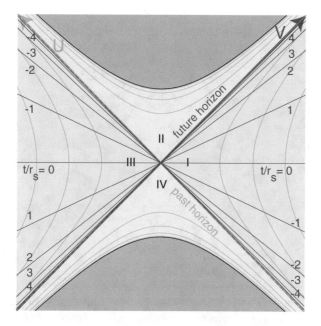

Fig. 6.9 A Schwarzschild black hole in light cone Kruskal–Szekeres coordinates. Lines of constant coordinate t are highlighted

Well, almost but not quite. To see what distinguishes regions III and IV from I and II, we look at the ratio U/V. According to (6.26) we have

$$\frac{U}{V} = -e^{\frac{-u-v}{2r_s}} = -e^{-\frac{t}{r_s}} \tag{6.33}$$

We use this relation to draw the lines of equal coordinate t in Fig. 6.9. Now remember that t was the time coordinate of an observer asymptotically far away from the horizon. In accordance with this picture, we see that in the region I (which we have identified as being outside the horizon) t gets larger as we go to the future. In region III however this is just the reverse: t decreases as we go to the future. If we compare region II to region IV we also see a clear difference there: While we cannot escape region II as we go into the future, we cannot go into region IV by going into the future. So in this sense regions III and IV are the time reverse of regions I and II respectively. For region III this means just flipping the sign of a coordinate, but region IV is the time reverse of a black hole, the white hole we mentioned in Sect. 6.1. It has an initial singularity at $r = 0$ (the second branch of the hyperbola $UV = 1$ in Fig. 6.8) and everything within is expelled in a finite amount of time. It is currently unknown and debated in the literature whether regions III and IV have any physical relevance at all. Nonetheless, the mathematical structure consisting of all four regions, which is called the maximally extended Schwarzschild solution, plays an important role in the understanding of static black hole solutions to the Einstein equation.

• **Observational Evidence for Black Holes:** Since classical black holes absorb everything, including light, they are rather difficult to detect directly. The first observational evidence for black holes was rather indirect. In 1971 Thomas Bolton, Louise Webster and Paul Murdin performed radial velocity measurements on the star HDE 226868, which was close to the bright X-ray source Cygnus X-1. From these measurements they concluded that there was an invisible companion star of more than 10 solar masses that was responsible for the X-rays. Since only a compact object can produce the required amount of X-rays and both a white dwarf and a neutron star could be ruled out based on the large mass, this constituted the first indirect observational hint of a black hole. The conclusion that a black hole was behind Cygnus X-1 was further strengthened when variability of the X-rays was discovered on the scale of 1 ms. This short term variability can only be explained if the source of the X-rays has an extent of no more than $c \times 1\,\mathrm{ms} \sim 300\,\mathrm{km}$—again suggesting that there is a black hole.

During the following decades, ever more precise radial velocity measurements in the cores of galaxies, especially the giant elliptic galaxy M87 and our milky way, found increasing evidence for very massive (million to billion solar masses), compact objects. Infrared images of the center of our galaxy, also called Sgr A^*, revealed a number of stars that orbit the galactic center in a matter of years.

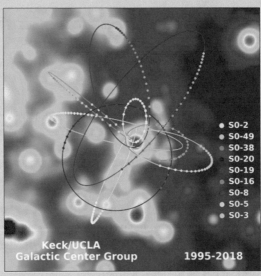

The image was created by Prof. Andrea Ghez and her research team at UCLA from data sets obtained with the W. M. Keck Telescopes.

From the dynamics of the stellar orbits, one can infer a mass of $\sim 4 \times 10^6\,M_\odot$ for Sgr A^*. The closest observed approach of a star to Sgr A^* of only

16 AU, which is closer than Uranus is to our Sun, leaves a black hole as the only reasonable explanation for the nature of Sgr A^*.

Today we have even more direct evidence for black holes by means of radio interferometry. The event horizon telescope collaboration, a network of radio telescopes in the millimeter range, spanning the entire globe, was able to directly image the shadow of the central black hole of M87.

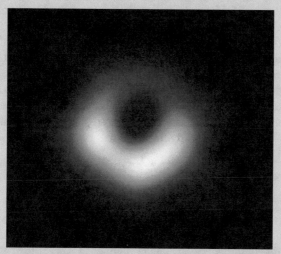

Image courtesy of the event horizon telescope collaboration.

How is it possible to obtain the required spatial resolution for such an image? Let us first compute the theoretical resolution limit of the interferometer. If we assume the telescope area spans the entire globe, the 'mirror diameter' is the diameter of the Earth ~ 13000 km. The observation was performed at a wave length $\lambda \sim 1.3$ mm, so the diffraction limit for the angular resolution $d \sim \lambda/D \sim 10^{-10}$ or about 2×10^{-5} arc s. Let us compare this to the angle subtended by the central black hole of M87. Since its mass is $\sim 7 \times 10^9 \, M_\odot$, we can estimate its size by the corresponding Schwarzschild radius $r_s \sim 2 \times 10^{10}$ km which is about 0.002 lt yr. The distance to M87 is about 54 million lt yr, so the angle subtended by one Schwarzschild radius of the central black hole M87* is about $\alpha \sim 0.002/(54 \times 10^6)$ or 7×10^{-6} arc s. What the image is showing is the accretion disk around the black hole and the central 'shadow'—the region where light does not escape anymore. The accretion disk does not extend all the way to the Schwarzschild radius, but only roughly to $3r_s$, the radius of the minimal stable circular orbit. We thus expect the diameter of the ring to be $\sim 6r_s$, which then corresponds to an angle of 4×10^{-5} arc s—just barely enough to be resolved by the event horizon telescope.

6.3 Formation of a Black Hole

Let us now come back to the question whether a black hole can form in some kind of realistic setting that does not include curiosities like initial white hole singularities. In fact, it is not too difficult to find a simple example of a gravitational collapse into a black hole. The best starting point are Eddington–Finkelstein coordinates (6.23), which have an interesting property: If we take the Schwarzschild radius $r_s = 0$ (which by the identity $r_s = 2GM$ means taking the mass of the black hole to zero) the Eddington–Finkelstein metric smoothly goes over into the metric of flat space. This gives us some hope that by interpolating between flat space and a black hole solution we can somehow describe the formation process of a black hole, and indeed we can. We picture some form of energy (which is called null dust) that is radially infalling at the speed of light into an empty, flat spacetime region. Since it is infalling at the speed of light, it travels along $v = $ const light rays and can be described by a function $m(v)$ if it has no angular dependence. Let us thus replace the constant $r_s = 2GM$ by the function $r_s = 2Gm(v)$. At early times we want m to vanish and at late times we want it to reach the full M, i.e.

$$m(v) \xrightarrow{v \to -\infty} 0 \qquad m(v) \xrightarrow{v \to \infty} M . \tag{6.34}$$

This is of course no more the Schwarzschild solution and we have to plug it into the Einstein equations to see what kind of energy momentum tensor can cause such a metric. Using one of the symbolic algebra routines in Appendix F, we see that there is only a single non-vanishing component of the Ricci tensor

$$R_{vv} = G_{vv} = \frac{2G}{r^2} \partial_v m(v) \tag{6.35}$$

which leads to an energy momentum tensor with the only non-vanishing component

$$T_{vv} = \frac{1}{4\pi r^2} \partial_v m(v) . \tag{6.36}$$

This specific, idealised form of matter is called Vaidya[6] null dust and the corresponding metric

$$d\tau^2 = \left(1 - \frac{2Gm(v)}{r}\right) dv^2 - 2dr\,dv - r^2(d\theta^2 + \sin^2\theta d\phi^2) . \tag{6.37}$$

is the Vaidya metric. It describes the formation of a black hole from spherically symmetric null dust without any coordinate singularity.

We introduce one further idealisation to make the core idea of black hole formation more intuitively understandable. Let us assume that the collapsing null dust forms

---

[6] Vaidya, Prahladbhai Chunilal, Indian physicist, *23.5.1918 Shahpur, †12.3.2010 Ahmedabad.

an infinitesimally thin shell so that

$$m(v) = M\Theta(v) \quad \text{and} \quad T_{vv} = \frac{M}{4\pi r^2}\delta(v) \,. \tag{6.38}$$

In this case, our spacetime is divided into two distinct regions. For $v < 0$, we have flat space while for $v > 0$ we have a Schwarzschild solution with $r_s = 2GM$. In the flat space region we have $r = r^*$ and thus the boundary $v = 0$ between the two regions in flat space coordinates is $t = -r$. If we are sitting in the center, at $r = 0$, spacetime is locally flat until at $t = 0$ the collapsing thin shell of null dust reaches us and a singularity forms. Before that however, at the time $t = -r_s = -2GM$, the collapsing thin shell has crossed its own Schwarzschild radius and a black hole with an event horizon at r_s has formed. No information about this reaches an internal observer before $t = 0$ and the formation of the black hole is thus nothing special for an observer inside the collapsing shell.

Of course at $t = 0$ there is a true, physical singularity at $r = 0$ that we cannot get rid of by any coordinate transformation. Whether this singularity gets regulated by quantum corrections or some other modification of the Einstein equations or whether it is sufficient to always shield it from the rest of the world by horizons is the topic of current research and beyond the scope of this book.

How does this formation of the black hole look from the perspective of an observer outside the shell at $r > r_s$? Since the thin shell is always at $v = 0$ and $v = t + r^*$, the shell at time t is located at $r^* = -t$. Using the relation between r and r^* (6.19), we see that viewed from the outside it takes a time

$$\begin{aligned} t_{r_0 \to r_1} &= -r_1 - r_s \ln\left(\frac{r_1}{r_s} - 1\right) + r_0 + r_s \ln\left(\frac{r_0}{r_s} - 1\right) \\ &= r_0 - r_1 + r_s \ln\frac{r_0 - r_s}{r_1 - r_s} \end{aligned} \tag{6.39}$$

for the shell to go from r_0 to r_1. This expression diverges as $r_1 \to r_s$, so from the perspective of the outside observer the black hole never even forms! The outside observer just sees the shell approach the Schwarzschild radius asymptotically and never crossing it. In this sense the answer to the question whether the black hole has even formed has become observer dependent. This is certainly one of the most extreme consequences of general relativity.

Before closing this chapter we would like to stress that what we have presented here barely scratches the surface of the vast area of black hole physics. We have not discussed charged or rotating black holes and we have not said anything about quantum effects at the horizon. The latter, in particular, have been puzzling researchers for almost half a century. Hawking has computed (see insert) that black holes decay via thermal radiation in a finite amount of time $T \propto M^3$ for an external observer. This implies that singularities cannot be shielded indefinitely by horizons and, among other things, has led to the suggestion that the horizon after all is of physical relevance and everything crossing it would burn up (the so-called firewall hypothesis). These topics are discussed controversially at the time of writing and one needs to keep these caveats in mind when applying classical black hole solutions of the Einstein equation to physical scenarios.

- **Perspective—Hawking Radiation:** While the Hawking effect is a genuine quantum effect that requires the apparatus of quantum field theory in curved spacetime for its derivation, we can get some glimpse of the mechanisms at work by a handwaving argument that uses minimal input from quantum physics. To obtain a first, qualitative picture of what is happening very close to the horizon, let us imagine that a quantum fluctuation produces a pair of particles, each having an energy E_0. Our input from quantum theory is that generically quantum fluctuation are suppressed exponentially in the action $S = Et$ by a factor $\propto \exp(-Et/\hbar)$. Applied to our specific case this means that a pair of particles, each possessing an energy E_0, are produced with a probability $\propto \exp(-2E_0 t/\hbar)$ and therefore are not very long lived. Close to the horizon however it can happen that out of the pair of particles produced one vanishes behind the horizon, while the other escapes off to infinity. Let us look at the fate of the two particles, starting with the escaping one. As it moves away from the black hole, it looses energy. If it is a photon, we can actually compute its redshift and thus derive the energy E as seen by an observer very far away from the horizon as

$$E = E_0 \sqrt{1 - \frac{r_s}{r}} \ . \tag{6.40}$$

For the infalling particle on the other hand we assume that it crosses the horizon in a time t. Once it is behind the horizon, there is no way back as we saw and thus the outgoing particle is 'produced' by the horizon of the black hole. So how long does it take for the infalling particle to cross the horizon? To answer this, we look at the near horizon coordinates where we have computed the proper distance ρ to the horizon (6.15). For a photon falling in perpendicularly we could just equate the horizon distance to the time t. If the photon is not perpendicular to the horizon it will take somewhat longer, but still the time it takes to cross the horizon will be proportional to the horizon distance. So generically we expect $t = \rho b$ with some proportionality constant $b > 1$.

Using this and the near horizon relation (6.15), we can write

$$E_0 t = E_0 \rho b \simeq \frac{E}{\sqrt{1 - \frac{r_s}{r}}} 2\sqrt{r_s(r - r_s)}b = 2E\sqrt{r_s r}b \simeq 2Ebr_s \ . \tag{6.41}$$

Remembering that the Schwarzschild radius was $r_s = 2GM$, we finally arrive at $E_0 t \simeq 4MbE$. With one particle swallowed by the black hole and the other escaping to infinity with an energy E, we can now interpret the probability of the quantum fluctuation

$$e^{-2E_0t} \simeq e^{-\frac{8GMbE}{\hbar}} \tag{6.42}$$

as a probability distribution for the energy E of the particles produced by the black hole. Interestingly, this is a Boltzmann distribution $\exp(-E/k_BT)$ corresponding to a temperature

$$T = \frac{\hbar}{8k_B GbM} . \tag{6.43}$$

We are thus led to the conclusion that black holes emit thermal radiation and are coincidentally black in the sense of a thermodynamic black body.

The reader might have noticed a certain discrepancy in our treatment of the outgoing and infalling particle. While we demanded that the infalling particle crosses the horizon in a time t, the outgoing one was allowed to escape to infinity and basically live forever. How can this be consistent? The apparent inconsistency is resolved by remembering that time dilation diverges as we approach the horizon. So while the outgoing particle lives forever in the frame of the distant observer, this time is squished to a tiny, finite amount for the infalling particle. For the distant observer on the other hand, the infalling particle is slowed down asymptotically as it approaches the horizon and one may say that the quantum fluctuation is 'frozen' there.

To proceed further and fix the constant b we need to account for quantum effects in a slightly more rigorous fashion. Remember that we started out by saying that quantum fluctuations are generically suppressed by a factor $\propto \exp(-Et/\hbar)$. This is somewhat of an oversimplification that has its roots in an idea by Feynman,[a] known as the path integral. While for classical systems the least action (or more precisely the extremal action) principle holds, Feynman showed that quantisation can be done by allowing all possible states, not just the least action one, and summing over them with an exponential weight factor that includes the action $S = Et$. The factor however is a complex phase $\propto \exp(iEt/\hbar)$, not an exponential suppression. In order to turn this into a Boltzmann factor, we would need an imaginary time t, or, a bit more precisely, we would need all the contributing states to only be valid in some sense on some finite imaginary time. But this is in fact what we have! Remember that following (6.26) we mentioned the curious mathematical fact that Kruskal–Szekeres coordinates U and V, and therefore the metric (6.32) of a black hole, are unchanged if one replaces t by $t + i4\pi r_s$? All states on this metric are thus periodic in imaginary time with a period $t = i4\pi r_s = i8\pi GM$. As t is the time coordinate for the observer at infinite distance we conclude that such an observer sees the black hole radiating off states, e.g. photons, of energy E with a probability

$$\propto e^{-\frac{8\pi GME}{\hbar}} ,$$ (6.44)

which again is thermal with a temperature

$$\boxed{T = \frac{\hbar c^3}{8\pi k_B GM}}$$ (6.45)

(here and in the remainder of this box c appears explicitly). This is in perfect agreement with our first, more intuitive derivation and fixes the unknown constant to $b = \pi$. The temperature T is the famous Hawking[b] temperature of a Schwarzschild black hole.[c] In SI units, we find that the Hawking temperature is

$$T \simeq 1.2 \cdot 10^{23} \text{K} \frac{1}{M[\text{kg}]} .$$ (6.46)

A black hole with the mass of a large asteroid has room temperature. The temperature of a black hole of one solar mass is a fraction of a $1\,\mu\text{K}$, which is significantly colder than the cosmic microwave background.

Our derivation is sketchy of course and is based on a semi-intuitive description of quantum effects. Done properly in quantum field theory, one for instance obtains the correct Bose–Einstein distribution instead of the classical Boltzmann distribution for photons.

If the black hole is radiating off energy, its own mass will decrease and, if there is no influx of energy, it should ultimately evaporate. We can make a crude estimate of this process by noting that the energy of the black hole is its mass $E = Mc^2$. If we further assume that only photons are radiated and that the black hole is a perfect black body with a surface area equal the area of the horizon $A = 4\pi r_s^2 = 16\pi G^2 E^2/c^8$, we can use the Stefan-Boltzmann law of thermodynamics, relating the energy emanating from a black body per unit time and surface area to the fourth power of the temperature

$$\frac{\mathrm{d}E}{A\mathrm{d}t} = \sigma_{SB} T^4 \qquad \sigma_{SB} = \frac{\pi^2 k_B^4}{60\hbar^3 c^2} ,$$ (6.47)

where σ_{SB} is the Stefan-Boltzmann constant. Inserting the Hawking temperature we obtain

$$\frac{\mathrm{d}E}{\mathrm{d}t} = \sigma_{SB} T^4 A = \frac{\hbar c^{10}}{15360\pi G^2 E^2}$$ (6.48)

for the energy radiated off. We therefore get a differential equation for the total energy of the black hole

$$E^2 \mathrm{d}E = -\frac{\hbar c^{10}}{15360\pi G^2}\mathrm{d}t , \tag{6.49}$$

which we can integrate assuming that the black hole starts out with an initial mass M_0 and after a time t it has evaporated. Thus

$$\int_{M_0}^0 M^2 \mathrm{d}M = -\frac{\hbar c^4}{15360\pi G^2}t \tag{6.50}$$

and we obtain for the lifetime of a black hole

$$t = \frac{5120\pi G^2}{\hbar c^4} M_0^3 . \tag{6.51}$$

---

[a]Feynman, Richard Philips, American physicist, *11.5.1918 New York City, USA, †15.2.1988 Los Angeles, USA. He received the Nobel prize in physics in 1965 for his contribution to the development of quantum electrodynamics.

[b]Hawking, Stephen William, British physicist and cosmologist, *8.1.1942 Oxford, UK, †14.3.2018 Cambridge, UK.

[c]The equivalence between periodicity in imaginary time and finite temperature might seem very ad-hoc here, but it is actually a well established result and forms the basis of thermal quantum field theory. It has applications far beyond Hawking radiation and the interested reader is referred to [13].

6.4 Problems

1. In this exercise we look at geodesics of objects that fall straight towards the center of the Schwarzschild metric ($\dot{\theta} = \dot{\varphi} = 0$).

   (a) Write down the equations of motion and solve them to obtain expressions for both $\mathrm{d}r/\mathrm{d}t$ and $\mathrm{d}r/\mathrm{d}\tau$.

   (b) Now concentrate on the special case where the radial velocity vanishes as $r \to \infty$ (this condition fixes the last remaining constant of motion). Find both the eigentime τ and the coordinate time t it takes to fall from $4r_s$ to r_s. What do these results imply for the time it takes for the infalling object to fall from $4r_s$ to r_s as seen from perspective of the object itself and from the perspective of a remote observer?

2. In this problem we will look at a static, radially symmetric solution to the Einstein equations that is not asymptotically flat for $r \to \infty$.

   (a) Show that the Schwarzschild-de Sitter metric

$$d\tau^2 = f(r)dt^2 + \frac{1}{f(r)}dr^2 + r^2 d\theta^2 + r^2 \sin^2 \theta d\varphi^2 \,,$$

where $f(r) = 1 - \frac{2MG}{r} - \frac{1}{3}\Lambda r^2$, is a solution to the vacuum Einstein equation

$$G_{\mu\nu} - \Lambda g_{\mu\nu} = 0$$

(b) Now let us assume that $MG\sqrt{\Lambda} \ll 1$. Does this metric have any horizons? Try to obtain their values approximately. *Hint: Start by assuming that the second term in f(r) dominates over the third term and compute the small correction due to the third term. Then assume conversely that the third term dominates and do the same.*

(c) Check whether this metric allows for circular (i.e. $r = $ const .) orbits.

3. The central black hole of our milky way galaxy, Sgr A*, has a mass of approximately 4 million solar masses. The supermassive black hole in the center of M87 has a mass of approximately 7 billion solar masses. Assuming that they do not gain any more mass (not even from the cosmic microwave background), how long will it take for them to decay? Under the same assumptions, how large a mass would a primordial black hole (i.e. one produced at the time of the Big Bang) have to have so that it would decay now?

Chapter 7
Basics of Modern Cosmology: Overview

The second part of this book is devoted to cosmology. In the following chapters the current standard cosmological model is developed via the combination of general relativity and astronomical observations—from the time of Hubble to the present day. In this chapter we anticipate its main results along the timeline of the evolution of the universe, which sets the general stage for what follows. We shall familiarize ourselves with the notion of an expanding universe—without considering the underlying physics at this point—and discuss a number of major consequences arising from this scenario.

7.1 History of the Universe

If we look into the night sky with our bare eyes or using a telescope, it is almost unimaginable that the universe could ever have been hot, dense and homogenous. But in the subsequent chapters we shall argue that this seems to be the best description of the early universe. If the matter content is constant, then the density of matter is proportional to a^{-3}, where a is some characteristic scale factor of the universe. We shall see in Sect. 10.1 that radiation is thinned out even more. In addition to the thinning out with increasing volume, its wavelength is stretched by a redshift factor $\propto a^{-1}$. Therefore the radiation energy density is proportional to a^{-4}. This implies that at earlier times radiation, which we can observe today in the form of the cosmic microwave background (CMB), must have played a prominent role. Remnants of the early epochs are imprinted on the fine structure of the CMB and in fact in the distribution of galaxies in our universe. Based on this information and our knowledge of physics at high energy scales, we can with some confidence extrapolate to times when the age of universe was only a tiny fraction of a second. If we push to the very beginning however, at some point our established knowledge fails. The time scales involved are so small and the energy densities are so high, that we cannot reproduce them in any experiment today. Here the frontiers of cosmology and high energy

© Springer Nature Switzerland AG 2020
R. Hentschke and C. Hölbling, *A Short Course in General Relativity and Cosmology*, Undergraduate Lecture Notes in Physics, https://doi.org/10.1007/978-3-030-46384-7_7

physics meet and we enter the realm of speculation with theories that at the moment are more or less reasonable guesses. Let us start at this point and sketch briefly what happened as the universe successively aged, cooled and expanded. Tables 7.1 and 7.2 as well as Fig. 7.1 help to relate temperature, processes, particles and their rest energies.

- Popular illustrations depicting the timeline of the universe, akin to our Fig. 7.1, typically begin with a point at $t = 0$ which is denoted as the Big Bang. In a box below we shall address this nomenclature in some detail—embedded in its historical context. However, it must be clear that we do not know anything about this 'point', what it is or how to describe it. A possible place were we have something to say, at least on dimensional grounds, and where we shall start is the Planck scale (cf. Table 7.1).
- The first $\sim 10^{-44}$ s are believed to be the realm of quantum gravity. There are only speculations about that period and it is even unclear whether time is well defined. According to the theory of inflation, the following $\sim 10^{-36}$ s are dominated by the vacuum energy of a scalar field, which led to an exponential expansion of the universe. At the end of this phase, the vacuum energy was transferred to other fields thereby heating up the universe.

Table 7.1 Thermal history of the universe. The first row in this table is the Planck scale, i.e. the time, t, is the Planck time t_P (cf. Appendix A). The attendant temperature is either $m_P c^2 / k_B$, where m_P is the Planck mass, or $h/(t_P k_B)$, i.e. 10^{32} K. The remainder of the table contains rough numbers following according to the proportionality $T \propto t^{-1/2}$ discussed below

Age (s)	Temperature (K)	Energy	Process (GeV)
10^{-44}	10^{32}	10^{19}	Quantum gravity
10^{-30}	10^{25}	10^{12}	Particle proc.
10^{-12}	10^{16}	10^{3}	Particle proc.
10^{-4}	10^{12}	10^{-1}	Particle proc.
10^{0}	10^{10}	10^{-3}	Nuclear proc.
10^{2}	10^{9}	10^{-4}	Nuclear proc.
10^{13}	$3 \cdot 10^{3}$	$3 \cdot 10^{-10}$	Atomic proc.

Table 7.2 Particle properties

Particle	Symbol	Rest energy (MeV)	$T_{threshold}$ 10^9 (K)
Photon	γ	0	0
Electron	e^-, e^+	0.511	5.930
Muon	μ^-, μ^+	105.7	1226
π-meson	π^0	134.9	1566
	π^+, π^-	139.6	1620
Proton	$p, \bar{p}$	938.3	10888
Neutron	$n, \bar{n}$	939.6	10903

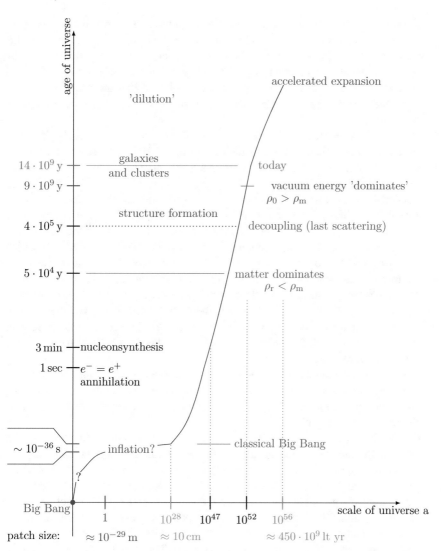

Fig. 7.1 History of our universe. Note that 'patch' refers to a patch of universe the size of one horizon distance at the beginning of inflation. What exactly a 'horizon distance' is will we discussed below. Here it is sufficient to state that is the distance over which two points are causally connected

- During the next $\sim 10^{-12}$ s the universe cooled down to a temperature of $\sim 10^{16}$ K, corresponding to an average particle energy of $\sim 10^3$ GeV. Present day experiments are unable to recreate these conditions and consequently this period is also the realm of speculative theories in particle physics such as supersymmetry, grand unification or extra dimensions. It is widely believed that the matter-antimatter asymmetry of our universe has its origin in this epoch and also the mysterious dark matter.

- At $\sim 10^{-12}$ s we enter the realm of experimentally accessible energies. High energy particle colliders, such as the large hadron collider (LHC), can probe these regions and our theoretical predictions are thus on a much more solid footing. For instance above ~ 100 GeV all known elementary particles are massless. This energy was reached when the early universe was $\sim 10^{-10}$ s old and had cooled to $\sim 10^{15}$ K. At this time, one of the fields present in the early universe, the Higgs field, acquired a vacuum expectation value, which caused all known elementary particles to obtain a mass (see box). The phenomenon is known as electroweak symmetry breaking and had profound consequences for the subsequent evolution of the universe. As the temperature was falling and the corresponding average kinetic energies of particles were dropping below their respective masses, they could no longer be pair produced out of the vacuum. Thus, the particles either decayed if they were unstable or became non-relativistic if they were stable. Electroweak symmetry breaking also separated the electromagnetic from the weak nuclear force. In fact, the weak force up to this point was not weak at all, but was approximately as strong as the electromagnetic force is today. Due to the electroweak symmetry breaking however, the carriers of the weak force became massive and as the universe cooled down further, they became increasingly difficult to produce, which is the reason the force is weak today.

- During the next $\sim 10^{-4}$ s the universe was a relatively homogenous plasma of particles interacting predominantly via the strong and electromagnetic force. As the average energy dropped below their pair creation energies, the heaviest of the particles, the top, bottom and charm quarks and the tau lepton vanished. Then, after the average energy fell below ~ 150 MeV, the universe entered into another phase, although technically speaking there was no phase transition but rather a smooth crossover. The temperature became too low to counteract the tendency of the strong nuclear force to confine quarks into bound states. Thus free quarks vanished from the universe and protons and neutrons were formed. Shortly thereafter muons could also no longer form.

- Until the universe was about one second old, neutrinos were in thermal equilibrium with the rest of the plasma. Neutrinos are very light particles and they only interact via the weak interaction. Because of this, their interaction with the remaining plasma became weaker as the universe cooled until they decoupled when the universe was about one second old. Today these neutrinos are redshifted and are believed to form a cosmic neutrino background possessing a temperature of ~ 1.95 K. Observation of this neutrino background would provide direct access to the first second of our universe's history. Thus far, however, the extremely weak interaction of these neutrinos with ordinary matter has prevented this.

- After the neutrinos decoupled, the remaining universe was an electromagnetic plasma of electrons, positrons, protons and neutrons. During the next few seconds the temperature fell below $\sim 10^9$ K, causing positrons and electrons to annihilate. This process heated the plasma and contributed to the current value of the CMB temperature, i.e. ≈ 2.725 K, which exceeds the expected neutrino background temperature.

- After a couple of minutes, our universe entered a phase during which two effects were competing and the result of this competition can be directly observed today. The two effects are the decay of the neutron, with a half-life of about 15 min and the synthesis of nuclei out of protons and neutrons that could only proceed once the temperature of the universe fell significantly below their binding energies. By chance this temperature was reached at this time as well. Because the chance of three or more nucleons meeting in one spot was negligible already, the synthesis of nuclei had to proceed through first forming deuterons, which have a relatively small binding energy of about 2.2 MeV. Once deuterons could be formed however, almost all of them, in total a mass fraction of $\sim 25\%$ of all baryonic matter, immediately combined to the much more stable ^{4}He nuclei with a small admixture of heavier elements and other isotopes. This process is known as Big Bang nucleosynthesis (BBN) and is one of the pillars of modern cosmology. Of particular interest is the fact that the abundance of trace elements (deuterium, ^{3}He and lithium) depends sensitively on the baryon density at the BBN epoch, which can in turn be inferred from the baryon density today. The consistency of the predicted abundances with observation is therefore a valuable crosscheck of the standard Big Bang scenario and in particular implies that whatever constitutes the bulk of dark matter today must have been non-baryonic already a few minutes after the Big Bang.

- For the next 400000 years the universe was a quite homogenous electromagnetic plasma of light nuclei with a background of decoupled neutrinos and supposedly dark matter, all the while it cooled and expanded. Initially photons were by far the dominant energy contribution, because protons and neutrons were only the remnant of a tiny symmetry breaking between matter and antimatter. As the universe expanded however, the electromagnetic radiation was redshifted and became progressively less important until at about 50000 years its energy density fell below that of non-relativistic matter. The universe continued to be a relatively homogeneous plasma until after ~ 400000 years its temperature fell below ~ 3000 K, allowing electrons and nuclei to combine into electrically neutral atoms. This process, confusingly referred to as recombination, led to the decoupling of the photons and the universe became transparent. The boundary separating the transparent from the previously opaque universe is known as the surface of last scattering. The immensely redshifted photons from this surface of last scattering form the cosmic microwave background (CMB) radiation which we observe today. Because it represents the boundary of the earlier opaque to the later transparent universe, it also denotes the limit of how far back in time we can see with optical telescopes.

- The universe after recombination still was very different from our present universe. For another few hundred million years, during which the CMB photons cooled down from ~ 3000 K to ~ 40 K, no stars or galaxies were formed yet and the universe was dark, which is somewhat poetically referred to as the dark ages. Once the first galaxies formed and the first stars lit up, the radiation could again ionize parts of the neutral gas, which is known as reionization. During the next ~10 billion years, the universe was matter dominated: it expanded and its expansion was decelerated by its matter content, whose density steadily decreased.

- Today we know, and we shall learn how to calculate this, that about 4 billion years ago the matter density of our universe, which now is roughly 14 billion years old, fell below that of the cosmological constant or dark energy. At this point, experiments as well as theoretical estimates tell us, the expansion of the universe started to accelerate and we appear to live in a dark energy dominated epoch.

- **What Do We Mean by Big Bang?** When the observational evidence of an expanding universe had solidified in the middle of the 20th century, there were essentially two theories proposed as an explanation. The first one was the steady state hypothesis, which conjectured that the average density of the universe would stay constant and thus matter would be constantly created throughout the expanding universe at a small rate. The alternative scenario, that matter was somehow created all in one go at the beginning of the universe, was given the name Big Bang by Fred Hoyle,[a] who himself developed the steady state theory [14]. Thus Big Bang referred rather generically to the hot, dense beginning of our universe.

Cosmological models have advanced significantly since then. Current theories distinguish between different epochs during the hot, dense beginning of our universe and thus we need to be more careful with what we mean by Big Bang. In astrophysical cosmology, Big Bang usually refers to the very beginning of the universe. Following this definition, all other events like nucleosynthesis or the hypothetical inflationary epoch occur after the Big Bang and are not part of it. Another sensible definition is that of new inflationary models which were invented to provide reasonable initial conditions for the observed, 'classical' expansion of the universe (cf. Chap. 11). According to this definition, the Big Bang is the starting point of the 'classical' expansion of the universe, i.e. the expansion as originally envisioned by the Big Bang theory. In this definition, the hypothetical inflationary epoch precedes the Big Bang and provides it with reasonable initial conditions.

In this book, when we generically refer to the 'Big Bang' we mean the usual 'beginning of the universe' definition. In those cases where we specifically mean the beginning of the classical expansion after inflation, we use the term 'classical Big Bang'.

---

[a]Hoyle, Fred, British astronomer, *24.6.1915 Gilstead, UK, †20.8.2001 Bournemouth, UK.

• **The Higgs Mechanism:** The Higgs[a] mechanism (also called Brout–Englert–Higgs mechanism) is a concept from particle physics that is extremely important for cosmology. It is the notion that elementary particles do not fundamentally possess mass, but that they acquire it when the universe cools down in a thermodynamic phase transition. How can we understand this?

Let us remember that in quantum mechanics a free spin-0 particle is described by the wave function

$$\psi(t, \vec{x}) = e^{-i\left(\frac{E}{\hbar}t - \frac{\vec{p}}{\hbar}\cdot\vec{x}\right)}.$$

If the free particle is relativistic, its wave function satisfies the Klein–Gordon equation

$$\left(-\hbar^2 \partial_\mu \partial^\mu - m^2\right)\psi(t, \vec{x}) = 0$$

which, according to the de Broglie relation, has the physical content

$$\left(E^2 - p^2 - m^2\right)\psi(t, \vec{x}) = 0.$$

Imagine now that our free particle is massless, i.e. the mass term is not there in the above equation. It is not entirely missing however, but instead it is replaced by an interaction term with a field ϕ, our Higgs field. We thus have

$$\left(-\hbar^2 \partial_\mu \partial^\mu - \phi^2\right)\psi(t, \vec{x}) = 0$$

and the dynamics of our free particle will depend on what this Higgs field does. Let us for simplicity imagine that ϕ is just a classical field, which lives in some potential

$$V(\phi) = \frac{1}{2}M^2\phi^2 + \frac{1}{4}\lambda\phi^4.$$

For $M^2 > 0$ (and $\lambda > 0$) this potential has a minimum at $\phi = 0$, while for negative M^2 the minimum is located at $\pm\phi_0 = \pm\sqrt{-M^2/\lambda}$ as shown in the following figure:

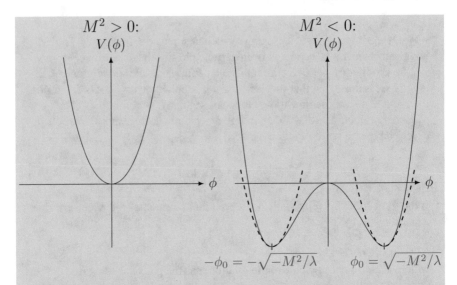

Suppose now that the classical field relaxes to its minimum value (roughly corresponding to the vacuum for a quantum field). In the case $M^2 > 0$, the result will just be that the ϕ field vanishes everywhere $\phi = 0$. Consequently, it will not produce any term in the Klein–Gordon equation

$$\left(E^2 - p^2\right)\psi(t, \vec{x}) = 0$$

which thus describes a massless particle. If we now lower M^2, the potential minimum at $\phi = 0$ will get shallower until we reach negative $M^2 < 0$ where suddenly we will have two minima at $\phi_0 = \pm\sqrt{-M^2/\lambda}$. Based on some tiny random fluctuations, the field will decide to fall into either of the two minima, but whichever one it chooses, when it settles into the minimum its value will be $\phi^2 = -M^2/\lambda$. Thus the Klein–Gordon equation now becomes

$$\left(E^2 - p^2 + M^2/\lambda\right)\psi(t, \vec{x}) = 0$$

so it has obtained a mass $m^2 = -M^2/\lambda$.

Spontaneous symmetry breaking, which here happens when the aforementioned random fluctuations 'select' one or the other minimum of the potential, is occurring in many physical systems. Perhaps the simplest example is a magnet heated above its Curie temperature. The elementary magnetic moments inside the material tend to align, but thermal fluctuations give rise to an order-to-disorder transition at the Curie temperature, beyond which the disordered phase, i.e. the phase of higher symmetry, has no net magnetization. Here $V(\phi)$ is replaced by $F(m)$, the free energy of the magnet in terms of its magnetization. When the temperature is reduced to the Curie temperature and below, i.e. into the ordered phase, random thermal fluctuations spontaneously drive the system into one or the other local minimum, reducing its free energy. This ordered phase is then characterized by either a positive or negative equilibrium magnetization.

Similarly, in quantum field theories the parameters of the theory, like the M^2 in our simple example, will depend on temperature. In the Standard Model of elementary particle physics, there is a scalar field, the Higgs field, which will develop a minimum away from $\phi = 0$, a so called vacuum expectation value, below a temperature of about 100 GeV. All the particles that are coupled to the Higgs field—which as far as we know includes all matter particles and the carriers of the weak nuclear force—thus obtained their mass when the universe cooled down below a temperature of about 100 GeV while they were massless as long as the universe was hotter.

---

[a] Higgs, Peter, British physicist, *29.5.1929 Newcastle upon Tyne. Nobel prize in physics 2013 together with François Englert.

7.2 Expanding Space and the Hubble Law of Recession

Let's get straight to the point, which is the expanding universe and its geometry. Figure 7.2 shows galaxies in a model one-dimensional space. The distance between any two galaxies is

$$D(\Delta\theta) = a\Delta\theta . \tag{7.1}$$

Note that the arc-shaped one-dimensional space in Fig. 7.2 suggests that a is the radius of a circle, i.e. (7.1) describes the spatial distance between two galaxies in terms of the product of the circle's radius times their angular distance. But the scientists[1] in our one-dimensional universe do not know this of course.

However, they do notice that their universe appears to be homogeneous.[2] This means that no matter in which of the galaxies they live the universe looks the same to them. One of the scientists makes a puzzling discovery. He notices, based on redshifted spectral lines in the starlight, that the galaxies to his left and to his right recede with a speed v proportional to their distance, i.e.

$$v = HD . \tag{7.2}$$

---

[1] We ignore here such trivial problems as whether galaxies or scientists can form in one dimension or how they can see past one another.

[2] And, in higher dimensions, isotropic.

Fig. 7.2 Galaxies in a
one-dimensional space

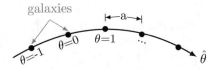

The proportionality constant is H. Thus, the greater the distance of a galaxy is from the observer the faster it moves away from him. The discovery is puzzling because of the above assertion of homogeneity. Other observers in any of the other galaxies should also obtain (7.2). But for us, having realised the true circular nature of the one-dimensional universe, this is not puzzling at all!

Imagine that a, which we shall call the scale factor[3] changes with time t, i.e.

$$a = a(t) . \tag{7.3}$$

For us, beings of a higher-dimensional world, it merely means that the radius of the circular universe changes for some reason or another. This in turn implies that $D(t) = a(t)\Delta\theta$ as well as $v(t) = \frac{d}{dt}D(t) \equiv \dot{D}(t)$ are functions of time. Inserting these expressions into (7.2) yields

$$\dot{a}(t) = Ha(t) . \tag{7.4}$$

In addition there is the possibility that H also is a function of time, i.e.

$$\frac{\dot{a}(t)}{a(t)} = H(t) . \tag{7.5}$$

During the lifetime of the observer this time-dependence of H may not be detectable however.

Equation (7.2) is called Hubble's law and H is called Hubble's constant in honour of Edwin Hubble[4] who deduced this from detailed measurements in the 1920s. Actually, the relation (7.2) was published about one year earlier by Georges Lemaître,[5] who also concluded the expansion of the universe, which Hubble initially didn't.

In the following we need the current value, which means the value H has today at time $t = t_0$, for most of our explicit calculations. The value is given in the last table of Appendix A:

$$\boxed{H_0 \equiv H(t_0) \approx 70\,\text{km s}^{-1}\,\text{Mpc}^{-1}} \tag{7.6}$$

---

[3]This quantity should not be confused with acceleration.

[4]Hubble, Edwin, American astronomer, *Marshfield 20.11.1889, †San Marino (California) 28.9.1953.

[5]Lemaître, Georges Edouard, catholic clergy, astronomer and cosmologist, *Charleroi 17.7.1894; †Löwen 20.6.1966.

or

$$\boxed{1/H_0 \approx 4.4 \cdot 10^{17}\,\text{s}} \tag{7.7}$$

Note that v can increase or decrease with time. For instance $a \propto t^2$ means that v increases. But for $a \propto \log t$ we find that v decreases. It was always expected that the expansion of the universe was decelerating because of the pull of gravity. We now know that this is not the case and that the universe is expanding at an accelerated rate. We discuss this in some detail in Chap. 10. Note also that v in (7.2) may exceed the speed of light! Distant galaxies may get out of touch. The velocity of distant galaxies can be measured by the redshift of the observed light. The distance is much harder to determine. Hubble's constant at first was too big by a factor of ten.

• **The Observed Value of the Hubble Constant:** From the first published value of the Hubble constant (by Georges Lemaître in 1927) throughout almost the entire 20th century, the numbers obtained were more order of magnitude than precise measurements. The main reason for this unsatisfactory state of affairs is the difficulty of precise distance measurements over cosmological scales. Direct, geometrical distance measurements are only possible within the small stellar neighbourhood within our own galaxy. Every extragalactic distance measurement is based on brightness measurements of 'standard candles'—astronomical objects such as for instance type Ia supernovae or certain types of variable stars—whose brightness is known so that their relative distances can be inferred. To measure cosmologically relevant distances, it takes a succession of several different standard candles, usually referred to as the 'cosmic distance ladder'. In the 21st century, this situation has improved dramatically. Distance measurements have become much more precise and alternative ways of measuring H_0 have been proposed so that its value can be obtained with percent level accuracy. In fact, at the time of writing there are several determinations with an accuracy of a few percent and they do not all agree. The most recent 'classical' measurement of H_0 uses the brightness of Cepheid variable stars and type Ia supernovae as steps in the cosmic distance ladder and obtains $H_0 = 74.03 \pm 1.42\,\text{km s}^{-1}\text{Mpc}^{-1}$ [15]. Substituting the most massive red giant stars instead of Cepheid variables results in a value $H_0 = 69.8 \pm 1.9\,\text{km s}^{-1}\text{Mpc}^{-1}$ [16].

Gravitational lenses offer an alternative path towards measuring H_0. Gravitational lenses may produce multiple images of the lensed (background) object. The length of the light paths differ among the different images and thus there is a time delay between the arrival of light rays that were emitted at the same time (see Problem 2 in Chap. 5). If the background object is variable on short time scales, the relative time delay can be measured and from it one can infer H_0. The most recent such measurement resulted in $H_0 = 73.3^{+1.7}_{-1.8}\,\text{km s}^{-1}\text{Mpc}^{-1}$

[17]. The currently most precise value for H_0 however is extracted from fluctuations in the cosmic microwave background radiation that we will discuss in Sect. 10.3. Without going into details we note that assuming the standard cosmological model a value of $H_0 = 67.4 \pm 0.5$ km s^{-1}Mpc^{-1} is extracted from CMB data [18]. Yet another independent determination is possible from Big Bang nucleosynthesis data in conjunction with data on galaxy distributions resulting in $H_0 = 68.3^{+1.1}_{-1.2}$ km s^{-1}Mpc^{-1} [19]. A number of additional independent methods to determine H_0 have been suggested, but none of them have a competitive precision (yet).

There currently is no agreement on whether the partially incompatible values of H_0 signify a problem with the standard cosmological model or will vanish over time as methods advance. For our purposes it is sufficient to note that $H_0 \sim 70$ km s^{-1} Mpc^{-1}.

Figure 7.3 depicts a possible time development of the circular universe with the spacetime metric

$$d\tau^2 = dt^2 - a^2(t)d\theta^2 . \tag{7.8}$$

The universe it describes is homogenous and bounded.

Fig. 7.3 A possible time history of the circular universe

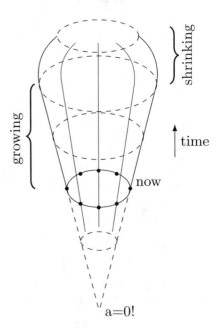

A generalisation to three-dimensional space and four-dimensional spacetime is

$$d\tau^2 = dt^2 - a^2(t)dx^2 - a^2(t)dy^2 - a^2(t)dz^2 .$$ (7.9)

Note that homogeneity means that a does not depend on position in space and isotropy means that a is the same in every direction. The assumption of both homogeneity and isotropy of the universe is called the cosmological principle. In a perfectly homogeneous and isotropic universe that has no other dynamics, the spatial coordinates of everything it contains stay fixed. The metric (7.9) therefore describes a very special reference frame, which is called comoving. More generally, we can rewrite (7.9) as

$$d\tau^2 = \left(1 - a^2(t)|\dot{\vec{r}}|^2\right) dt^2$$ (7.10)

and define an object as comoving if $\dot{\vec{r}} = 0$. One may also wonder why the coefficient in front of dt^2 in (7.9) is unity. However, any coefficient different from one, e.g. $f(t)dt^2$, may be absorbed into a new definition of time, i.e. $dg(t)^2 = f(t)dt^2$. Obviously, the time dependence of the scale factor is of great significance and this is what we want to study next.

7.3 Problems

1. Galaxies typically possess peculiar velocities of a few hundred kilometers per second. In order to distinguish these velocities from the speed of recession between galaxies they must have a sufficiently large separation. How large should their distance be?

2. There are comoving observers in two galaxies separated by the proper distance δl. A particle passes the first observer with velocity v en route to the second galaxy and the second observer. What is the dependence of the particle's momentum p on the scale factor a? Assume that $v \gg \delta u$, where δu is the velocity of the second observer relative to the first.

Chapter 8
Friedmann–Robertson–Walker Cosmology

If we average over sufficiently large scales—on the order of 100 Mpc—our universe is homogeneous and isotropic to a high degree. We can therefore study what the Einstein equations tell us about the scale factor $a(t)$ of the FLRW metric. But before we do this, we want to find out what we can infer about the time dependence of $a(t)$ from ordinary Newtonian dynamics.

8.1 Newtonian Friedmann Equation

Figure 8.1 depicts a patch of universe containing galaxies that is well described by Newtonian dynamics. All masses move with speeds far less than the speed of light and the curvature radius is much bigger than the size of the patch. Let us take the perspective that we are in the center. Before we study this picture in detail, we want to investigate the motion of ordinary particles in the field of a heavy mass in Newtonian gravity. This situation is illustrated in Fig. 8.2. The small particle with mass m moves radially at a distance D with respect to the large mass M. Energy conservation demands

$$\frac{1}{2}mv^2 - \frac{GmM}{D} = E .$$ (8.1)

The sketch on the right in Fig. 8.2 shows trajectories of m when it is ejected away from M by an initial thrust. In the case $E < 0$ the small mass remains bound to the large mass, i.e. the initial velocity is smaller than the required escape velocity. When $E \geq 0$ the initial velocity is equal or greater than the escape velocity and the small mass keeps moving away from M forever.

Note that the universe behaves in the same way and its equation of motion has the same property. In order to see this, let us return to Alice and Bob. Bob has meanwhile

© Springer Nature Switzerland AG 2020

R. Hentschke and C. Hölbling, *A Short Course in General Relativity and Cosmology*, Undergraduate Lecture Notes in Physics,
https://doi.org/10.1007/978-3-030-46384-7_8

Fig. 8.1 Uniform patch of
universe, including galaxies,
with us at its center

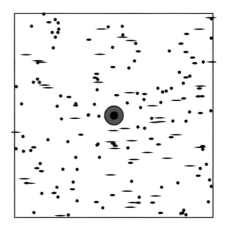

Fig. 8.2 Ordinary particle,
m, moving in the
gravitational field of a heavy
mass, M

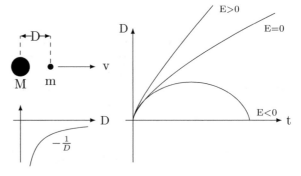

stopped accelerating and reached galaxy B, which he declares to be the center of the
universe, at the coordinate origin $x = 0$. Alice has remained in galaxy A at $x \neq 0$,
as depicted in Fig. 8.3. According to the Hubble law (7.1), the physical distance and
the coordinate distance between the two are related via

$$D = x\, a(t) .\qquad(8.2)$$

Their relative velocity is

$$v = \dot{D} = x\, \dot{a}(t) ,\qquad(8.3)$$

if neither of them moves with respect to the (expanding) coordinate system. Inserting
this into the energy conservation condition (8.1) yields

$$x^2 \dot{a}^2(t) - 2\frac{GM}{x\, a(t)} = \frac{2E}{m} .\qquad(8.4)$$

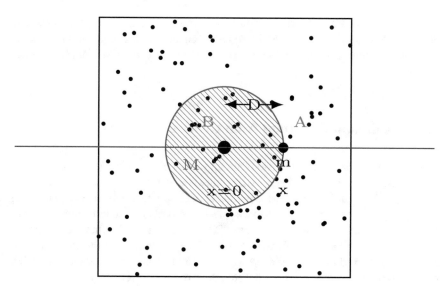

Fig. 8.3 Bob's and Alice's galaxies

Here M is the entire mass inside the red sphere in Fig. 8.3. We can express this mass in terms of the uniform but time dependent mass density

$$\rho(t) = \frac{M}{4\pi x^3 a^3(t)/3} \tag{8.5}$$

and thus we obtain

$$x^2 \dot{a}^2(t) - \frac{8\pi}{3} \rho(t) G x^2 a^2(t) = \frac{2E}{m} . \tag{8.6}$$

Although E is conserved, its actual value depends on an arbitrary choice, namely the decision of Bob to center the coordinate system at his position. But since Alice knows about the cosmological principle, she can decide to place the coordinate origin anywhere else. If she does E changes but the quantity $K = -2x^{-2}E/m$ is unaffected, because

$$\left(\frac{\dot{a}(t)}{a(t)}\right)^2 = \frac{8\pi G}{3} \rho(t) - \frac{K}{a^2(t)} , \tag{8.7}$$

which is a coordinate independent equation for the scale factor $a(t)$ of the universe. Equation (8.7) known as the first Friedmann[1] equation. Obviously, the right hand side gives the square of the Hubble 'constant', which we can see will not be constant in general.

8.2 Friedmann Equations from General Relativity

Spheres in any number of dimensions are uniform, i.e. they possess constant curvature. Let us briefly discuss spheres in the context of cosmology, which is the cosmology of a spherical universe. If one is talking about a spherical universe one is talking about the geometry of space—not spacetime. Afterwards we can talk about the other alternative, which is negative curvature. There also is a space which is uniformly negative curved, which is harder to visualise. Then we talk about the cosmology of a negatively curved universe. Finally we also want to connect the equations, which we have written down so far, with the equations of general relativity.

Let us begin with d-dimensional spheres. A 1-sphere is a circle (embedded in two dimensions) as shown in Fig. 8.4. Its equation is

$$x^2 + y^2 = 1 \tag{8.8}$$

(unit circle). Our name for this geometry is Ω_1. A 2-sphere (embedded in three dimensions) is our ordinary sphere or rather its surface as shown in Fig. 8.5. Its equation is

$$x^2 + y^2 + z^2 = 1 \tag{8.9}$$

Fig. 8.4 A 1-sphere

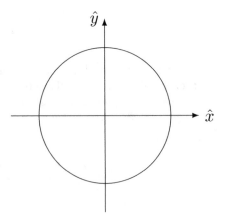

[1]Friedmann, Alexander Alexandrovich, Russian physicist, *Saint Petersburg 16.6.1888, †Leningrad 16.9.1925.

Fig. 8.5 A 2-sphere

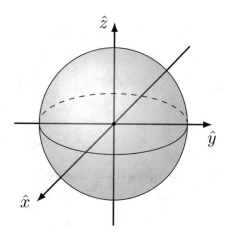

Fig. 8.6 A 1-sphere in internal coordinates

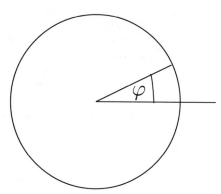

(unit sphere). Our name for this geometry is Ω_2. A (unit) 3-sphere has the equation

$$x^2 + y^2 + z^2 + w^2 = 1 . \tag{8.10}$$

Our name for this geometry is Ω_3.

Another way to do this is by describing spheres in terms of a metric, i.e. we use internal coordinates. The 1-sphere is described via

$$ds^2 = d\varphi^2 \equiv d\Omega_1^2 \tag{8.11}$$

(cf. Fig. 8.6). The 2-sphere is described via

$$ds^2 = d\theta^2 + \underbrace{\sin^2 \theta \, d\varphi^2}_{=d\Omega_1^2} \equiv d\Omega_2^2 \tag{8.12}$$

(cf. Fig. 8.7). By analogy we obtain for the 3-sphere

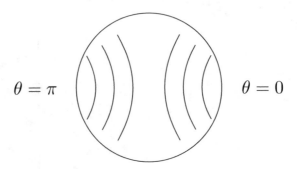

Fig. 8.7 A 2-sphere in internal coordinates

$$ds^2 = d\alpha^2 + \sin^2\alpha \, d\Omega_2^2 \equiv d\Omega_3^2 \; . \tag{8.13}$$

Here α is a new angle.

Suppose now that we want to make a d-sphere of radius a. All we must do is to multiply $d\Omega_d^2$ by a, i.e.

$$ds^2 = a^2 d\Omega_d^2 \; . \tag{8.14}$$

Now let's think about a cosmology where space is a 3-sphere. We are interested in spacetime and we want the radius of our sphere to change with time. The proper time here becomes

$$d\tau^2 = dt^2 - a^2(t) \, d\Omega_3^2 \; . \tag{8.15}$$

It is convenient to place ourselves at the pole (cf. Fig. 8.8). Looking out into space we find that the angular width of objects is greater than what we find for the same object at the same distance in flat space.

Now let's spend some time with a geometry which has uniform negative curvature. It's related to the metric of a sphere in a very simple way. Instead of

$$ds^2 = d\alpha^2 + \underline{\sin^2\alpha} \, d\Omega_{d-1}^2 \equiv d\Omega_d^2 \tag{8.16}$$

we now have

$$ds^2 = d\alpha^2 + \underline{\sinh^2\alpha} \, d\Omega_{d-1}^2 \equiv d\mathcal{H}_d^2 \; . \tag{8.17}$$

While in (8.16) $0 \leq \alpha \leq \pi$, in (8.17) $0 \leq \alpha \leq \infty$. Looking out into a $\mathcal{H}_3$-space we do see that distant objects appear smaller than they appear in flat space (cf. Fig. 8.9). The attendant spacetime is

$$d\tau^2 = dt^2 - a^2(t) d\mathcal{H}_3^2 \; . \tag{8.18}$$

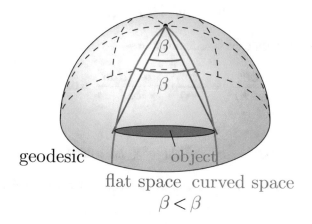

Fig. 8.8 Observational cosmology where space is a 3-sphere

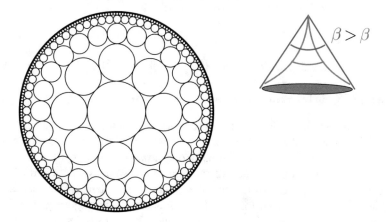

Fig. 8.9 Looking out into a space with uniform negative curvature

We can show that the metric (8.15) yields the Friedmann equation with $K = +1$, whereas the metric (8.18) yields the Friedmann equation with $K = -1$. Both are in fact special cases of the so called Robertson–Walker[2] (also referred to as the Friedmann–Lemaître–Robertson–Walker or FLRW) metric, which we write down in its standard form as

$$d\tau^2 = dt^2 - a^2(t) \left[\frac{dr^2}{1 - Kr^2} + r^2 d\Omega^2 \right], \tag{8.19}$$

where

---

[2]Robertson, Howard Percy, American mathematician and physicist, *27.1.1903 Hoquiam, †26.8.1961 Pasadena; Walker, Arthur Geoffrey, British mathematician, *17.7.1909 Waltford, †31.3.2001 West Chiltington.

$$d\Omega^2 = d\theta^2 + \sin^2\theta \, d\varphi^2 \ . \tag{8.20}$$

Setting $K = 1$ and using the substitution $r = \sin\alpha$ yields

$$\frac{dr^2}{1 - Kr^2} = \frac{dr^2}{1 - r^2} = \frac{\cos^2\alpha \, d\alpha^2}{\cos^2\alpha} = d\alpha^2 \ . \tag{8.21}$$

Hence

$$d\tau^2 = dt^2 - a^2(t)\left(d\alpha^2 + \sin^2\alpha \, d\Omega_2^2\right) \ , \tag{8.22}$$

which is (8.15). Setting $K = -1$ and using this time the substitution $r = \sinh\alpha$ yields

$$\frac{dr^2}{1 - Kr^2} = \frac{dr^2}{1 + r^2} = \frac{\cosh^2\alpha \, d\alpha^2}{\cosh^2\alpha} = d\alpha^2 \ . \tag{8.23}$$

Hence

$$d\tau^2 = dt^2 - a^2(t)\left(d\alpha^2 + \sinh^2\alpha \, d\Omega_2^2\right) \ , \tag{8.24}$$

which is (8.18). In between the two cases is $K = 0$, which is flat space.

Remark The fact that the curvature constant K only has the discrete values 0 or ± 1 might seem to imply that the three cases are completely distinct without one being the limit of another. This impression is wrong however and $K \in \{0, \pm 1\}$ is just a convenient choice. It implies that the coordinate r is dimensionless and the scale factor $a(t)$ has dimensions of length—the radius of the sphere in the case $K = 1$. We could instead have opted for K to have units of inverse length squared. In that case, $a(t)$ is dimensionless, the radial coordinate r has units of length as usual and it is easy to see that the curvature radius is given by $\sqrt{|K|}$ at a time when the scale factor is $a(t) = 1$. In that setup we can smoothly vary K to have arbitrary positive or negative values which correspond to positive or negative curvature with the limiting flat case $K = 0$ in between.

We can work out the Einstein tensor for the above FLRW metric. You can find this calculation in Appendix F. In particular for the 00-element we obtain

$$\mathcal{G}_{00} = \frac{3\left(K + \dot{a}(t)^2\right)}{a(t)^2} \ . \tag{8.25}$$

Hence from (4.37) (with $\Lambda = 0$) and $\mathcal{T}_{00} = \rho$

$$\boxed{\left(\frac{\dot{a}(t)}{a(t)}\right)^2 = \frac{8\pi G\rho(t)}{3} - \frac{K}{a^2(t)}} \ , \tag{8.26}$$

which agrees with our previous result for the first Friedmann equation (8.7).

There is yet another equation, which we obtain if we work out the Ricci scalar, i.e.

$$R = -2\mathcal{G}_{00} - 6\frac{\ddot{a}(t)}{a(t)} \tag{8.27}$$

(cf. Appendix F). Tracing over the Einstein equation (4.37) (again with $\Lambda = 0$) we see that $R = -8\pi G T^\mu_\mu$ (cf. (4.34)) and we obtain

$$\frac{\ddot{a}(t)}{a(t)} = -\frac{4\pi G}{3}\left(2T^0_0 - T^\mu_\mu\right) . \tag{8.28}$$

We now have to find the trace of the energy momentum tensor T^μ_μ for the cosmic medium. Since it is invariant under coordinate transformations, we evaluate it in the most convenient reference frame, which is the comoving frame. Because we assume our universe to be homogeneous and isotropic, the spatial diagonal (pressure) components are just $T^i_i = -3P$ where P is the pressure.[3] We already know that $T^0_0 = \rho$, so the second Friedmann equation reads

$$\boxed{\frac{\ddot{a}(t)}{a(t)} = -\frac{4\pi G}{3}(\rho + 3P)} . \tag{8.30}$$

Repeating the steps leading to the Friedmann equations, this time using (4.37) with $\Lambda \neq 0$, we find

$$\left(\frac{\dot{a}(t)}{a(t)}\right)^2 = \frac{8\pi G\rho(t)}{3} - \frac{K}{a^2(t)} + \frac{\Lambda}{3} \tag{8.31}$$

and

$$\frac{\ddot{a}(t)}{a(t)} = -\frac{4\pi G}{3}(\rho + 3P) + \frac{\Lambda}{3} . \tag{8.32}$$

With the definitions of the energy density ρ_v and the attendant pressure P_v via

$$\boxed{\rho_v = -P_v = \frac{\Lambda}{8\pi G}} , \tag{8.33}$$

---

[3] Note that

$$T^{\mu\nu} = (\rho + P)u^\mu u^\nu - Pg^{\mu\nu} , \tag{8.29}$$

where u^μ is the four-velocity (cf. [12]).

we once again obtain (8.26) and (8.30) if we consider ρ_v a part of ρ and P_v a part of P. In other words, the cosmological constant can be absorbed into ρ and P. In P it causes a negative (partial) pressure. The reason why we use the subscript v for vacuum instead of the subscript Λ will become clear in the next chapter when we discuss the different components of the overall energy density of the universe.

Let us now combine (8.30) with the time-derivative of the first Friedmann equation (8.26). We obtain

$$\boxed{\dot{\rho}a^3 + 3(\rho + P)a^2\dot{a} = 0} , \tag{8.34}$$

which is energy conservation! It is most striking that ρ_v and P_v cancel each other in this expression. Furthermore, this equation, like the first Friedmann equation, can be derived in an entirely different fashion, this time from thermodynamics. Energy conservation in thermodynamics is encoded in the first law which we write as

$$dE = -PdV , \tag{8.35}$$

where E is the internal energy. We do not include TdS, i.e. the entropy contribution, on the right hand side, because we assume the expansion of the universe to be adiabatic.[4] By expressing E as well as V in terms of the scale factor and the energy density we find

$$d(a^3\rho) = -Pda^3 \tag{8.36}$$

or

$$a^3 d\rho + 3(\rho + P)a^2 da = 0 . \tag{8.37}$$

'Division' by dt yields (8.34).

8.3 Problems

1. (a) Consider a two-dimensional world, which in three dimensions is the surface of a sphere. What is the relation between the width δl of an object in the 2D world and its distance $r(\gg \delta l)$ from the 2D observer, the attendant angular width $\delta\varphi$ and the radius R of the sphere?
   (b) Imagine that many of the above objects are arranged in a closed circle of radius r around the observer. Calculate the circumference S of the circle. What is the accuracy the observer needs if he wants to distinguish his curved

---

[4]Adiabaticity of the expansion of the universe has in fact been an implicit assumption that we have made very early on. Without it we would not be able to absorb the full time dependence of the metric into the scale factor!

space from flat space assuming that in his world the circle has a radius of 1000 m and R = 6400 km (roughly the radius of the Earth).

(c) Thus far the width of the object was measured along a small circle,where it has the angular width $\delta\varphi$. More appropriate is the measurement along a great circle, where it has the angular width $\delta\phi$. Show that $1 - \cos\phi = \sin^2\theta\,(1 - \cos\varphi)$ and thus $\delta\phi = \sin\theta\,\delta\varphi$

2. (a) Find the cartesian coordinates $x = x(\varphi, \theta, \alpha)$, $y = y(\varphi, \theta, \alpha)$, $z = z(\theta, \alpha)$, $w = w(\alpha)$ which yield $d\Omega_3^2$ in (8.13). *Hint: In the case of $d\Omega_3^2$ the answer is $x = \cos\varphi\,\sin\theta$, etc.*

(b) Find the coordinates $x = x(\varphi, \alpha)$, etc. corresponding to $\mathcal{H}_2$. Make a sketch of the attendant geometrical shape in a x, y, z-coordinate system.

(c) The physical distance element on the surface of a 2-sphere of radius α in a three-dimensional space with constant positive curvature $(K = 1)$ is

$$ds^2 = a^2 \sin^2\alpha\,d\Omega_2^2$$

Calculate the surface A of the sphere as well as its volume. What is the volume in the limit of small α and for $\alpha = \pi$.

(d) Repeat part (c) (except for the limit $\alpha = \pi$) for constant negative curvature $(K = -1)$, where

$$ds^2 = a^2 \sinh^2\alpha\,d\Omega_2^2 \ .$$

3. Work out the components of the Einstein tensor from the FLRW metric.

4. (a) For $P > -\rho/3$ show that a closed universe collapses after reaching a maximum size. Flat and open universes, however, expand forever.

(b) Discuss the behavior of $a(t)$ for $-\rho \le P \le -\rho/3$.

(c) What happens if $-\rho > P$?

5. Derive the relation

$$\dot{H} = -4\pi G(\rho + P) + \frac{K}{a^2} \ .$$

6. Assuming a non-vanishing cosmological constant, show that the static solution for a universe filled with cold matter is unstable.

7. Using an approach analogous the classical 'derivation' of the first Friedmann equation, show that in the classical limit $\Delta\varphi = -3\ddot{a}(t)/a(t)$.

Chapter 9
Thermodynamics of the Universe

In the previous chapter, we have established the Friedmann equations as the evolution equations for the scale factor. To further understand the past and the future of our universe, we need to explore the relation between the scale factor and the individual components of the energy density. In addition we must tie the latter to the attendant partial pressures via suitable equations of state $P = P(\rho)$. This is the program of the current chapter.

9.1 The Temperature of the Universe

In 1989 a spectacular experiment was performed, measuring the cosmic background radiation spectrum to high precision [20]. The expected frequency dependence, intensity per frequency interval $d\omega$, should follow $\omega^3/(\exp[\beta\hbar\omega] - 1)$ or, if converted to wavelength, i.e. intensity per wavelength interval $d\lambda$, $\lambda^{-5}/(\exp[\beta hc/\lambda] - 1)$. The cosmic background radiation spectrum was found to be in complete agreement with Planck's[1] prediction of the radiation spectrum of a black body, implying that the early universe must have been in thermal equilibrium. The radiation spectrum measured by COBE is shown in Fig. 9.1, comparing the measured data to the theory at a background radiation temperature of 2.725 K.

It is important to note that the perfect agreement seen here was not easily obtained. Previous attempts to measure the spectrum using balloons or ground based antennas had provided results even suggesting that the microwave background doesn't follow the Planck formula (cf. for instance Chap. 4 in [21]). It is not surprising that COBE

---

[1] Planck, Max, German physicist, *Kiel 23.4.1858, †Göttingen 4.10.1947; his solution of the black body problem laid the foundation to the development of quantum theory. He received the Nobel prize in physics in 1918.

© Springer Nature Switzerland AG 2020
R. Hentschke and C. Hölbling, *A Short Course in General Relativity and Cosmology*, Undergraduate Lecture Notes in Physics, https://doi.org/10.1007/978-3-030-46384-7_9

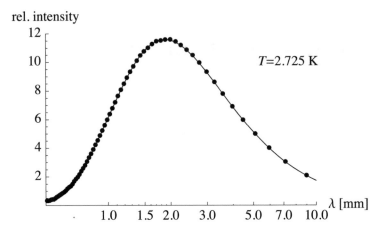

Fig. 9.1 Cosmic background radiation intensity and Planck's prediction versus wavelength. Data points are from the FIRAS instrument on the COBE satellite

was only the first in a series of satellites (COBE, WMAP (Wilkinson Microwave Anisotropy Probe), Planck, ...) dedicated to the exploration of the microwave background. After all, the microwave background is perhaps the most important source of information for current cosmologists.

The cosmic background radiation, when it was first (officially[2]) detected in 1964 by Arno Penzias and Robert W. Wilson[3] was not unexpected. Already in the 1940s scientists had pondered the origin of heavy elements in the cosmos. It was clear that there must have been an earlier epoch of the universe with substantially higher temperatures and densities to synthesise the heavier elements we find today. In 1948 R. Alpher, G. Gamow and R. Herman developed a theory of the cosmic origin based on a hot initial state. They also estimated the abundance of helium and predicted the cosmic background radiation and its Planck spectrum. Quite independently but certainly inspiringly Georges Lemaître, who can be called the father of the Big Bang scenario, explained the recession of the galaxies within the framework of general relativity (1927)[4] and considered the 'atomic size' origin of the universe (1931).

We have briefly mentioned in Sect. 7.1 that the current temperature of the CMB is about 3 K and that it consists of the redshifted photons from the surface of last scattering when the universe had a temperature of about 3000 K. Why this temperature is what it is and why it is defined rather sharply is discussed in Appendix C. Here we want to understand how this temperature evolved as the universe expanded.

---

[2]The radiation was observed (but not recognized as CMB) more than 20 years earlier in the rotation spectra of CN molecules in the interstellar medium.

[3]Both researchers received the 1978 Nobel prize in physics for their discovery.

[4]Expanding spaces in this context were studied even earlier by de Sitter.

9.2   Matter, Radiation and Vacuum Energy

Let us start by investigating more closely the first Friedmann equation (8.26). It relates the square of the expansion rate of the universe $H(t) = \dot{a}(t)/a(t)$ to the energy content and curvature. Because it does not contain the pressure, we do not need the relation between energy density and pressure, the equation of state, yet. On the other hand, we only get the square of the Hubble constant, so this first Friedmann equation does not tell us if we have expansion or contraction.

What about the curvature term? We certainly know that in our vicinity space is flat. Even if we look as far back in time as we can, i.e. to the surface of last scattering—the CMB, we see no traces of a spatial curvature of our universe.[5] The curvature radius of our universe, which in our convention is given by the scale factor $a(t)$, is therefore much larger than the patch of universe visible to us. Consequently, $r \ll 1$ in the FLRW metric (8.19) and a flat universe $K = 0$ will be a very good approximation at least in the patch of universe that is visible to us. Let us therefore explore the consequences of (8.26) for the case $K = 0$. Assuming that the universe does not shrink but expand we have

$$\frac{\dot{a}}{a} = \sqrt{\frac{8\pi G \rho}{3}} . \tag{9.1}$$

9.2.1   Matter and Radiation

In a universe with constant matter content the density of matter ρ_m is proportional to a^{-3}, i.e.

$$\rho_m \propto a(t)^{-3} . \tag{9.2}$$

The universe also contains a number of photons. If this number is constant, it means that the photon number density is proportional to a^{-3}, just like the matter density. To get the energy density of radiation ρ_r, we have to multiply the aforementioned number density by the photon energy h/λ. Here h is Planck's constant and λ is the photon wavelength. The photon wavelength, however, is not constant. In an adiabatically expanding universe it will stretch with the scale factor, i.e. $\lambda \propto a(t)$, and thus

$$\rho_r \propto a(t)^{-4} . \tag{9.3}$$

The stretching of the photon wavelength in an expanding universe is called redshift. It is important to note that the proportionality of λ and a is required if we want

---

[5]The idea is as follows: We look at the surface of last scattering, i.e. the microwave background, and determine the angular width, α, of the largest fluctuation patches. We know the distance d to the surface of last scattering and we know s, the actual size of the largest fluctuation patches, because we know how the fluctuations developed. Therefore we can determine the curvature. E.g., $s = d \alpha$ yields zero curvature.

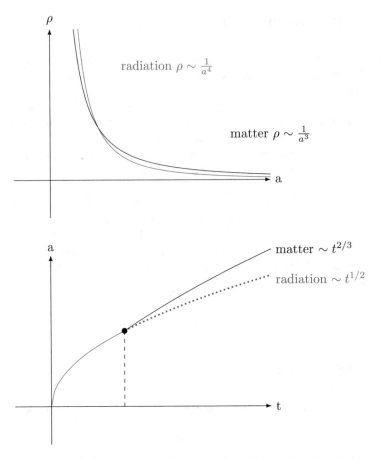

Fig. 9.2 Matter and radiation energy versus the scale factor, $a(t)$, as well as the scale factor versus time, t

consistency of ρ_r with Stefan's law of black body radiation (cf. the discussion in the context of (9.10)).

Solving (9.1) for $\rho = \rho_m$ as well as for $\rho = \rho_r$ we find that

$$a(t) \propto t^{\frac{2}{3}} \quad \text{matter dominates} \tag{9.4}$$

$$a(t) \propto t^{\frac{1}{2}} \quad \text{radiation dominates} \tag{9.5}$$

Figure 9.2 shows for both cases the energy density versus the scale factor as well as the scale factor versus time. It is evident that the radiation density decreased faster since the beginning of the universe. The cross over occurred when the universe was on the order of 10^4 years old, which we shall learn in the next chapter (and also later calculate).

It is interesting to note that for both a matter and a radiation dominated universe the Hubble constant

$$H(t) = \frac{\dot{a}(t)}{a(t)} = \begin{cases} \frac{2}{3}t^{-1} & \text{matter dominates} \\ \\ \frac{1}{2}t^{-1} & \text{radiation dominates} \end{cases} \tag{9.6}$$

is inversely proportional to the age of the universe t. Thus, the age of the universe can be inferred from measuring the Hubble constant (and knowing the correct cosmological model as we shall see).

Remark The combination of the two equations $D = x\,a(t)$ and $v = x\,\dot{a}(t)$ in conjunction with the limit that $v = 1$, i.e. the velocity of light, defines a (time dependent) horizon distance

$$D_h = H(t)^{-1} . \tag{9.7}$$

This is the distance over which two points are causally connected. Comparing this equation to the previous may look slightly confusing, but this is because we have agreed to set $c = 1$! The concept of a horizon distance will be important in the Chaps. 10 and 11. We note that (9.7) defines D_h in a somewhat handwaving fashion. An in depth discussion and a more precise definition of D_h is included in Chap. 10.

9.2.2 Vacuum Energy

The third possibility, a universe dominated by a cosmological constant or vacuum energy, is a constant energy density $\rho_v = \Lambda/8\pi G$ (cf. (8.33)). If this term dominates, the solution to (9.1) is exponential and we obtain

$$a(t) \propto e^{\sqrt{\frac{8\pi G \rho_v}{3}}t} . \tag{9.8}$$

This behavior is quite distinct from either the matter or the radiation dominated universe.

The distinguishing feature of the cosmological constant, when it is interpreted as the energy density of the vacuum, is its independence of the scale factor $a(t)$. When the universe expands it remains constant. We can see how counterintuitive this is in the nice and frequently used example of the homogenous and isotropic expansion of the universe, the balloon. When the ballon is inflated, its surface, corresponding to vacuum space in the universe, expands. However, the increase of the ballon's surface area is compensated by the decreasing thickness of the latex it is made of. Empty space is different. It does not become 'thinner' in the sense that vacuum energy diminishes. There is just more of the same as space expands.

But why should the vacuum possess an energy density? The answer is due to quantum effects. Even the simplest quantum mechanical systems, like the harmonic oscillator, have a ground state energy which is not zero. In the case of the harmonic oscillator it is $E_0 = \hbar\omega/2$. Similarly, quantum field theories, the currently best descriptions of all matter and radiation in the universe, have ground state energy. Formally this ground state energy is even infinite, but it is widely believed today that quantum field theories are a good description of nature only at distances larger than the Planck length $l_P = \sqrt{\hbar G}$ and energies below the Planck energy $E_P = \sqrt{\hbar/G}$. One would therefore generically expect vacuum energy densities of the order of one Planck energy per Planck length cube or $\rho_P \sim 1/(\hbar G^2)$, which in SI units is $\rho_P \sim 4.6 \cdot 10^{113}$ J/m^3! At this energy density, it would take the universe only $T_P \sim \sqrt{3\hbar G/(8\pi)} \sim 10^{-44}$ s to expand by a factor e, which is obviously not the case today.

The smallness of the vacuum energy as compared to its 'natural' value ρ_p in a quantum field theory is the most striking example of what is called a fine tuning problem in contemporary physics. In a sense this problem was deepened by the discovery, which we will discuss in detail in the next chapter, that the cosmological constant today is not exactly zero but possesses a value which corresponds to a vacuum energy density of $\rho_v \sim 5.3 \cdot 10^{-10}$ J/m^3 or an expansion time scale T of 17 billion years. Before that discovery one could have hoped that there was some natural mechanism—like a yet undiscovered symmetry—that requires the vacuum energy to vanish. A mechanism that suppresses it by more than 122 orders of magnitude without making it zero is far harder to come by.

On the other hand, a large vacuum energy density and the corresponding rapid, exponential expansion may be useful in understanding the very early development of our universe. In a conjectured inflationary phase the vacuum energy could have been far larger and thus the early universe could have expanded dramatically in a very short time. We will further explore this in Chap. 11.

9.3   Thermal History of Our Universe

Let us now try to link the expansion of the universe to the temperature of the radiation it contains. Our starting point is Stefan's law, which states that the radiation energy density in the universe is proportional to T_r^4, where T_r is the radiation temperature.[6] Remembering that $\rho_r \propto a^{-4}$ we thus have

$$T_r \sim a^{-1} . \tag{9.10}$$

---

[6]We can see this also via the following argument. In equilibrium thermodynamics we have

$$\left.\frac{\partial E}{\partial V}\right|_T = T\left.\frac{\partial P}{\partial T}\right|_V - P \tag{9.9}$$

But in the matter dominated era (9.4) holds and thus

$$T_r \sim t^{-2/3} . \tag{9.11}$$

This in turn implies

$$\frac{t_{today}^{2/3}}{t_{recomb}^{2/3}} \approx \frac{T_{recomb}}{T_{today}} \approx 1000 . \tag{9.12}$$

and using $t_{today} \approx 14 \cdot 10^9$ y with the assumption that the expansion of the universe was matter dominated we find

$$t_{recomb} \approx 4 \cdot 10^5 \text{ y} \tag{9.13}$$

for the recombination epoch.[7]

What does this mean for an astronomer with a good telescope? The situation is illustrated in Fig. 9.3. The figure shows planes in different colours. These planes correspond to the universe at different times. We neglect expansion of the universe to keep the picture simple (cf. the remark at the end of this section). The vertical black line is the time line of the astronomer (at rest). The bottom plane (blue) is the universe at the time of the Big Bang. The next plane (black) is the universe at the time of recombination. The red and the green planes are universes at latter times. At any time the astronomer can see only photons on his past light cone. In the red universe these photon paths are indicated by the red arrows. The 'oldest' photons are emitted from the red wavy circle in the black plane. This is when the universe became transparent. The wavy circle is the surface of last scattering, which in our case is a sphere rather than a circle. Today the astronomer is in the green universe. As before the oldest photons he observes are emitted from the green wavy circle in the black plane. Thus, as times passes the surface of last scattering grows in size. Note that the waviness of the surfaces of last scattering is meant to illustrate temperature fluctuations in the microwave background.

In Fig. 9.3 there are two points A and B along the green surface of last scattering. Note that a photon emitted at B at the time of the Big Bang can just barely reach A by the time of recombination. If we move A further away from B no information can be passed from B to A in that time. This also means that no thermal equilibration is possible between A and B. And yet, the microwave background is almost isotropic. This is the essence of the cosmological horizon problem, one of the main motivations

---

(e.g. [22]; Sect. 2.2). Here E is the internal energy and P is the pressure. Replacing the left hand side with ρ_r (E is extensive), and using $\rho_r = 3P_r$, which follows from (8.37) because $\rho_r \propto a^{-4}$, immediately yields $P_r \sim T_r^4$ or $\rho_r \sim T_r^4$.

[7] Let's assume we had not known that the radiation temperature today is about 3 K. We could have used the last line in Table 7.1 to guess the temperature at this time and then use (9.11) to calculate today's radiation temperature. The result is close to the actual value. What this really suggests is that the Planck scale is not a bad starting point.

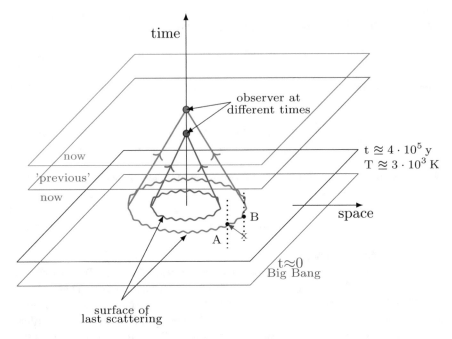

Fig. 9.3 Surfaces of last scattering

for the theory of inflation of the universe. In inflationary models, the universe begins sufficiently small to allow for equilibration. Inflation then stretches this small region to a size large enough to include the visible universe.

Remark Let us briefly come back to the statement 'We neglect expansion of the universe to keep the picture simple'. In order for Fig. 9.3 to be correct 'time' must be replaced by $\eta \equiv \int \mathrm{d}t/a(t)$, also called the conformal time, and 'space' by the coordinate radius. We shall encounter these quantities at the beginning of the next section. Their effect on Fig. 9.3 will be the subject of a problem in the context of inflation.

9.4   The Equation of State

In the present standard model of cosmology there are basically three components, compiled in Table 9.1, to the energy density on the right hand side of the Friedmann equation, i.e. (8.26):

- *matter*—particles or objects (galaxies, black holes, ...) at rest, i.e. moving with the overall expansion.
- *radiation*—photons whose wavelength grows with the expansion as $a(t)$.

Table 9.1 Components to the energy density

	Matter	Radiation	Vacuum energy
ρ	$\rho_m = \frac{\text{const}}{a^3}$	$\rho_r = \frac{\text{const}}{a^4}$	$\rho_v = \rho_0$
a	$\sim t^{2/3}$	$\sim t^{1/2}$	$\sim e^{\sqrt{\dots}t}$
pressure	$P_m = 0$	$P_r = \frac{1}{3}\rho_r$	$P_v = -\rho_0$

- *vacuum energy*—or (most likely) dark energy or cosmological constant

Let us now focus on the last line in Table 9.1. We expect the universe to possess a pressure P, which we try to express via

$$\boxed{P = \omega\rho} \tag{9.14}$$

Equation (9.14) is the cosmological equation of state.[8] We find ω using the first law of thermodynamics, which means energy conservation. Inserting (9.14) into the energy conservation equation (8.37) yields

$$\frac{d\rho}{\rho} = -3(1+\omega)\frac{da}{a} \tag{9.15}$$

and integration results in

$$\rho = \frac{\text{const}}{a^{3(\omega+1)}} . \tag{9.16}$$

For vacuum energy, we recover $\omega = -1$ as required by (8.33). For radiation we obtain $\omega = 1/3$, while $\omega = 0$ is implied for nonrelativistic (comoving) matter. Table 9.1 summarises these results.

Let us briefly comment on the 'pressureless dust' approximation $P_m = 0$ for matter. Our statement '(galaxies, black holes, ...) at rest, i.e. moving with the overall expansion', at the beginning of this section appears to be a physically satisfying motivation for $P_m = 0$. But there where times before there were galaxies etc. In addition, most matter appears to be of an unknown type. So maybe we should explore the possibility $P_m > 0$. However, before we can do this, we must continue for a while and collect additional information on the abundance of matter and the thermodynamic role of the radiation density.

A pictorial summary of the three components of 'matter' in the universe is given in Fig. 9.4. Note that the three different energy densities dominate in different eras depending on the scale factor a. The lower panel shows a sketch of a versus time t. A wavy line 'past which we cannot see' indicates the surface of last scattering. In the

---

[8]Note that the equation of state is not included in Einstein's field equations and must be added separately.

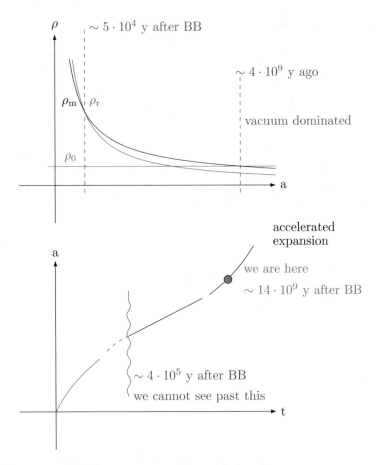

Fig. 9.4 Summary of the three components of 'matter' in the universe

upper panel the numbers are obtained based on the following approximate relative abundance of the different components at the present time, i.e.

$$\frac{\rho_m}{\rho_{\text{total}}} \approx 0.3 \quad (0.25 \text{ dark matter} + 0.05 \text{ baryonic matter})$$
$$\frac{\rho_r}{\rho_{\text{total}}} \approx 0 \tag{9.17}$$
$$\frac{\rho_v}{\rho_{\text{total}}} \approx 0.7$$

Note that 'baryonic matter' here refers to all matter that is mainly composed of baryons, like atoms or neutron stars. Within the Standard Model of particle physics, baryons are composite particles made of three quarks, the most common of which are protons and neutrons. Electrons e.g. are leptons, not baryons, but they are only referred to as non-baryonic matter in their free form. When bound in atoms they fall under baryonic matter.[9]

9.5 The Quantities Ω

Let us slightly rewrite the first Friedmann equation (8.26) as

$$\frac{3H_0^2}{8\pi G} = \rho - \frac{3K}{8\pi G a_0^2} \, , \tag{9.18}$$

where a_0 is the current scale factor. The quantity on the left hand side is an energy density. We call it the critical energy density at the current time

$$\rho_{c,0} = \frac{3H_0^2}{8\pi G} \tag{9.19}$$

and its relation to the total energy density ρ determines the overall spatial shape of the universe. For $\rho = \rho_{c,0}$ the universe is flat. For $\rho > \rho_{c,0}$ the universe has positive spatial curvature and for $\rho < \rho_{c,0}$ it has negative spatial curvature. In both cases[10] the curvature radius is

$$a_0^2 = \frac{3}{8\pi G |\rho - \rho_{c,0}|} \, . \tag{9.20}$$

The observed value of the critical density is

$$\rho_{c,0} \approx 8 \cdot 10^{-10} \text{ J/m}^3 \tag{9.21}$$

and since $c = 1$ we can also write it as

$$\boxed{\rho_{c,0} \approx 9 \cdot 10^{-27} \text{ kg/m}^3} \, , \tag{9.22}$$

which facilitates the intuitive understanding in terms of density of nonrelativistic matter. The critical density of the universe approximately corresponds to the mass of five hydrogen atoms per cubic meter.

Now we want to disentangle the different contributions to the energy density ρ at the present time. For matter, radiation and vacuum energy we define

$$\rho_{m,0} = \Omega_m \rho_{c,0} \quad \rho_{r,0} = \Omega_r \rho_{c,0} \quad \rho_{v,0} = \Omega_v \rho_{c,0} \tag{9.23}$$

---

[9]This slight discrepancy in nomenclature parallels the 'metallicity' in astrophysics which subsumes all elements that are not hydrogen or helium.

[10]Specifically $K = \pm 1$.

and also for curvature we can define an effective energy density via

$$\Omega_K = -\frac{K}{a_0^2 H_0^2} \, .$$ (9.24)

Therefore the first Friedmann equation at the present time simply reads

$$\boxed{\Omega_v + \Omega_m + \Omega_r + \Omega_K = 1} \, .$$ (9.25)

For all other times we know how each of the components scales with a and thus we can rewrite the Friedmann equation as

$$\left(\frac{H}{H_0}\right)^2 = \Omega_v + \Omega_m \left(\frac{a_0}{a}\right)^3 + \Omega_r \left(\frac{a_0}{a}\right)^4 + \Omega_K \left(\frac{a_0}{a}\right)^2 \, .$$ (9.26)

• **Example—Curvature Density:** We have seen that we can reinterpret the cosmological constant as an energy and pressure density. Equations (9.24) and (9.26) suggest that we can even treat the curvature term in the same way. From (8.26) we readily obtain the effective 'curvature energy density' as

$$\rho_K(t) = -\frac{3K}{8\pi G a(t)^2} \, .$$ (9.27)

Note that curvature density is thinned out less than ordinary matter as the universe expands, but more than vacuum energy. Consequently, if the universe is in a matter or radiation dominated phase the relative importance of curvature density increases with time but in a universe dominated by vacuum energy, it decreases. We will come back to this in Chap. 11 as an important hint for inflation.

Using either (9.16) or the observation that the curvature contribution is cancelled out in the second Friedman equation (8.30), we can also derive the equation of state

$$P_K = \omega_K \rho_K \qquad \omega_K = -\frac{1}{3} \, .$$ (9.28)

The exotic nature of curvature is clearly exhibited as either energy or pressure are negative (for either positive or negative curvature).

• **Example—A Universe from Nothing?** Consider the following idea. What if the sum of the individual energies of all elementary particles in the universe is equal to their mutual gravitational potential energy, i.e. the total energy of the universe is zero?

The plausibility of this idea can be demonstrated as follows. Let the gravitational potential energy of a test particle of mass m be

$$U = -\frac{GMm}{D} . \qquad (9.29)$$

Here $M = \frac{4\pi}{3} D^3 \rho_m$ and D represent the suitably defined mass and effective distance of the rest of the matter in the universe. Thus

$$U = -\Omega_m \left(\frac{H_0 D}{c}\right)^2 \frac{mc^2}{2} . \qquad (9.30)$$

Based on the above we would identify D with the horizon distance and thus expect $\mathcal{O}(H_0 D/c) = 1$, i.e.

$$U = -\mathcal{O}(mc^2) \qquad (9.31)$$

(where we include c explicitly). This means that the rest energy of the test mass and its gravitational potential energy could add up to zero.

A more extensive discussion of this idea can be found in [23].

9.6 Radiation Density Before the Matter Era

The equation of state (9.14) with $\omega = 1/3$, according to thermodynamics, leads to Stefan's law, i.e.

$$\rho_r = \sigma(k_B T)^4 \qquad (9.32)$$

(cf. [22]; p. 35. See also the discussion following (9.16).). In the same reference it is shown that in the case of photons, which are described by this equation of state, the factor σ is given by

$$\sigma_\gamma = g_\gamma \frac{\pi^2}{30(\hbar c)^3} , \qquad (9.33)$$

where $g_\gamma = 2$. Here $g_\gamma = 2$ accounts for the two spin states of the photon.

In the ultra-relativistic limit, i.e. the rest mass is negligible, this equation of state is believed to be generally valid for particles as well (cf. Sect. 112 in [12]). This means that (9.32) also holds. The remaining differences are expressed in terms of different g-factors. In the case of neutrinos this means

$$g_\nu = \underbrace{\frac{7}{8}}_{\text{fermion}} \times \underbrace{3}_{\text{types}} \times \underbrace{2}_{\nu+\bar\nu} \times \underbrace{1}_{\text{spin states}} = \frac{21}{4} . \tag{9.34}$$

In the case of electron-positron pairs

$$g_{e^-e^+} = \underbrace{\frac{7}{8}}_{\text{fermion}} \times \underbrace{1}_{\text{types}} \times \underbrace{2}_{e^-+e^+} \times \underbrace{2}_{\text{spin states}} = \frac{7}{2} \tag{9.35}$$

(cf. [24]; Table 1).[11] Thus, before the era of matter dominance in Fig. 9.4, the radiation density was given by the sum of the photon and the neutrino contributions, i.e.

$$\rho_r(T) = \Big[2 + \underbrace{\frac{21}{4}\Big(\frac{4}{11}\Big)^{4/3}}_{\approx 1.36}\Big] \frac{\pi^2 (k_B T)^4}{30(\hbar c)^3} . \tag{9.36}$$

The factor $(4/11)^{4/3}$ requires an explanation. When the temperature dropped below $5 \cdot 10^9$ K and the electron-positron pairs vanished, their entropy didn't vanish. Because at this time the neutrinos were not in thermal equilibrium with either photons or particles, this entropy was 'inherited' by the photons. Note that the entropy in the case of $P = \rho/3$ is given by

$$S \propto g T^3 \tag{9.37}$$

(cf. [22]; p. 35). Assuming the above adiabatic process[12] therefore means

---

[11] The factor $7/8$ arises due to the exchange of the usual bosonic integral

$$\int_0^\infty \frac{x^3 dx}{e^x - 1} = \frac{\pi^4}{15}$$

by the corresponding fermionic integral

$$\int_0^\infty \frac{x^3 dx}{e^x + 1} = \frac{7}{8}\frac{\pi^4}{15}$$

in the derivation of the black body radiation energy density.

[12] The word *adiabatic* can mean different things. Thus far we have used the term *adiabatic* for processes for which $\delta q = 0$, i.e. the system of interest is thermally insulated. Here we mean $dS = 0$. This is not quite the same though. The mathematical form of the second law of thermodynamics states $dS \geq \delta q/T$. Thus, $\delta q = 0$ generally means $dS \geq 0$! Only for a reversible process does $\delta q = 0$ also imply $dS = 0$.

$$g_{e^-e^+} T_{\text{before}}^3 + g_\gamma T_{\text{before}}^3 = g_\gamma T_{\text{after}}^3 . \tag{9.38}$$

Hence

$$T_{\text{before}} = \left(\frac{4}{11}\right)^{1/3} T_{\text{after}} . \tag{9.39}$$

It is believed that the neutrinos continued with T_{before} instead with T_{after}, which we usually just call T. It means that since this time the neutrinos have a lower temperature than the photons, which is accounted for by the factor $(4/11)^{4/3}$.

If we now insert $T = 2.7$ K into (9.36) we find

$$\rho_{r,0} \approx 7 \cdot 10^{-14} \text{ J/m}^3 \tag{9.40}$$

or

$$\boxed{\rho_{r,0} \approx 7.8 \cdot 10^{-31} \text{ kg/m}^3} \tag{9.41}$$

and therefore $\rho_{r,0}/\rho_{c,0} \approx 9 \cdot 10^{-5}$, i.e.

$$\boxed{\Omega_r \approx 9 \cdot 10^{-5}} . \tag{9.42}$$

• **Example—Neutrino Mass and Dark Matter:** We now know that neutrinos possess mass[a] (The most recent experiment designed to measure absolute neutrino masses is The KArlsruhe TRItium Neutrino (KATRIN) experiment (https://www.katrin.kit.edu)). Let's assume we know the number density n_ν of all neutrinos in the cosmic background. The (average) mass (in kg) of one such neutrino, assuming that the cosmic neutrinos account for all dark matter, follows according to

$$m_{\text{DM},\nu} = 0.25 \, \rho_{c,0}/(n_\nu c^2) . \tag{9.43}$$

Note that n_ν is given by

$$n_\nu = 3 \cdot \frac{3}{4} \cdot \frac{4}{11} n_\gamma . \tag{9.44}$$

Here n_γ is the background photon number density. The calculation of n_γ is very much the same as that of the photon energy density except for a factor $\hbar\omega$, the photon energy, which we have to omit. The result is

$$n_\gamma = 2 \cdot \frac{\zeta(3)}{\pi^2} \left(\frac{k_B T}{\hbar c}\right)^3 \approx 4.1 \cdot 10^8 \text{ m}^{-3} , \tag{9.45}$$

where $T = 2.725$ K and $\zeta(3) = 1.20206$ ($\zeta(s) = \sum_{k=1}^{\infty} k^{-s}$). The leading factor 2 accounts for the two polarization states. The factor 3 in (9.44) represents the three neutrino species. The next factor, $3/4$, arises due to the replacement of the bosonic integral, $\int_0^{\infty} x^2/(e^x - 1)\mathrm{d}x = 2\zeta(3)$, by its Fermion counterpart $\int_0^{\infty} x^2/(e^x + 1)\mathrm{d}x = (3/4)2\zeta(3)$. Finally, the factor $4/11$ is again the result of (9.39). Thus

$$m_{\mathrm{DM},\nu} = 6.7 \cdot 10^{-33} \text{ g} \quad \text{or} \quad \approx 3.8 \text{ eV} . \tag{9.46}$$

Note that this mass roughly corresponds to a temperature of 40000 K. Even though today the cosmic background temperature is much lower, it does not affect our calculation if we assume that the ratio $n_\nu/n_\gamma = 9/11$ has remained constant since the time when the neutrinos were in fact ultra-relativistic (For more details see Sect. 5.5 in [25].). Unfortunately, 3.8 eV, small as it may seem for an average neutrino mass, is firmly ruled out by experiment. At the time of writing, there is an upper bound of ~ 1 eV on one neutrino species by the aforementioned KATRIN experiment and neutrino oscillations constrain the mass differences between the species to a small fraction of an eV. What is not ruled out by these considerations is the possibility of a hypothetical fourth, much heavier neutrino species, which is usually termed 'sterile' because it is not associated to any of the three known lepton families of the Standard Model of particle physics. If it exists, such a neutrino could (and in fact must) be far more massive. We return to neutrinos as dark matter candidates in Sect. D.4.

[a]Nevertheless, with the exception of this box, we treat neutrinos as if they are massless.

We now have all the numbers which we need to calculate the fraction of baryonic matter already stated in (9.17). The present baryonic number density is given by the product of n_γ from (9.45) with the baryon-to-photon ratio η from (C.17). Hence an estimate for the baryonic energy density $\rho_{b,0}$ at the present time is

$$\rho_{b,0} \approx n_\gamma \eta m_p \approx 4 \cdot 10^{-28} \text{ kg/m}^3 , \tag{9.47}$$

where m_p is the proton mass. Therefore the current baryonic matter fraction is

$$\boxed{\frac{\rho_{b,0}}{\rho_{c,0}} \approx 0.05} . \tag{9.48}$$

It appears that roughly half of the baryons are not luminous matter, e.g. stars. In the context of the non visible baryons one often reads about the 'missing baryon problem'. On the other hand, experts do have a lot of confidence in the above baryonic matter fraction, because it is based on the detailed calculations of Big Bang nucleosynthesis (BBN) (cf. Appendix C) and it is in accord with the power spectrum of the cosmic microwave background (CMB).

• $P = 0$ **Versus the Ideal Gas Law** $PV = NT > 0$ (throughout this box we set the Boltzmann constant equal to unity): We have motivated $P = 0$ with the lumpy masses suspended in space. But there was a time when the content of the universe could be described as a 'fairly hot soup' consisting of non-relativistic baryons in equilibrium with the photons (prior to or right around recombination). Shouldn't the particles under these conditions possess a pressure greater than zero—like a gas at a certain density and temperature?

What is ρ_m for these particles? Note that the trace of the energy-momentum tensor is given by

$$T_i^i = \rho - 3P \quad \text{and} \quad T_i^i = \sum_\alpha m_\alpha \sqrt{1 - v_\alpha^2} \, \delta(\vec{r} - \vec{r}_\alpha) \,, \tag{9.49}$$

where α runs over the particles (cf. Sect. 34 in [12]). Because $v_\alpha \ll 1$ we express T_i^i by the leading terms of a Taylor-expansion, i.e.

$$T_i^i \approx \sum_\alpha m_\alpha \delta(\vec{r} - \vec{r}_\alpha) - \sum_\alpha \frac{1}{2} m_\alpha v_\alpha^2 \delta(\vec{r} - \vec{r}_\alpha) = \rho_{m,o} - \frac{3}{2} \frac{NT}{V} \,. \tag{9.50}$$

The first term is the rest mass density of the particles and the second term their thermal kinetic energy density. According to (9.49), where we set $\rho = \rho_m$,

$$\rho_m - 3P \approx \rho_{m,o} - \frac{3}{2} \frac{NT}{V} \tag{9.51}$$

and using the ideal gas law we find

$$\rho_m \approx \rho_{m,o} - \frac{3}{2} \frac{NT}{V} + \frac{6}{2} \frac{NT}{V} = \rho_{m,o} + \frac{3}{2} \frac{NT}{V} \,. \tag{9.52}$$

Note that N is the particle number.

Let's extend this by adding in the photon contribution, assuming equilibrium between particles and photons. We obtain for the total internal energy E

$$E = V\rho \approx V\rho_{m,o} + \frac{3}{2} NT + \sigma V T^4 \,. \tag{9.53}$$

According to the first law (energy conservation)

$$dE = d(mN + \frac{3}{2} NT + \sigma V T^4) = -(P_m + P_r) dV \,. \tag{9.54}$$

Here m is the particle mass (for simplicity we consider one type of particle only) and P_m and P_r are the partial pressures of the particles and the photons, respectively. Using $P_m V = NT$ and $P_r = \frac{1}{3}\sigma T^4$ as well as $V \propto a^3$, we obtain after some algebra (Problem 4)

$$d \ln T = -Y(x) d \ln a \quad \text{where} \quad Y(x) = \frac{1 + 3x/4}{1 + 3x/8}. \tag{9.55}$$

Here $x = \frac{N}{V\sigma T^3} = \frac{1}{3} P_m / P_r$. Currently we use $P_m = 0$. Thus $x = 0$ and $Y(x) = 1$, which yields $T \propto a^{-1}$—our previous result! But if $P_m \gg P_r$ then $Y(x \to \infty) = 2$ and consequently $T \propto a^{-2}$.

Can $P_m \gg P_r$ occur? Using $N \approx n_\gamma \eta$ it is not difficult to see that

$$\frac{P_m}{P_r} \approx \eta. \tag{9.56}$$

This result is independent of temperature and with $\eta \sim 10^{-9}$ we find that $P_m = 0$ is indeed a very good approximation. However, in order to convince yourself that the average universe is quite different from our usual surroundings, you should compute P_m / P_r inside an average office (Problem 3).

9.7 Problems

1. The radiation energy density for photons in Sect. 9.6 is $\rho_\gamma = \sigma_\gamma (k_B T)^4$. The energy per unit time emanating from the surface area A of a black body is

$$\frac{dE}{dt} = \frac{c}{4} \rho_\gamma A.$$

Note that $c\rho_\gamma$ is the total flux density from a surface volume element and $\sigma_{SB} = (c/4)\sigma_\gamma k_B^4 \approx 5.67 \cdot 10^{-8}$ W m^{-2} K^{-4} is the Stefan-Boltzmann constant.
Imagine a spherical volume of water possessing an initial temperature $T_i = 300$ K and a mass of 1000 kg.

   (a) Ignoring evaporation, what is its temperature after ten hours if the sphere is embedded in vacuum at 0 K. Remark: Use the heat capacity for water $C \approx 4.2$ kJ (kg K)$^{-1}$.
   (b) What would be the temperature increase of the water if it absorbes all impinging CMB photons during this time without giving off any energy.

2. In an infinite static universe containing stars at a constant number density, the (night) sky should be absolutely bright. Its temperature should be the surface temperature of the stars. The argument in favour of this so called Olbers' paradox

is as follows. Note that the radiation intensity from a star decreases as R^{-2}, where R is the distance from the star to the observer. However, the number of stars grows as R^3 and therefore the overall intensity diverges as R goes to infinity. There are numerous reasons why this is not observed, e.g. the universe is neither infinite nor static, nor is the life of a star infinite, the expansion of the universe leads to horizons beyond which the light of a star cannot ever be seen, etc. But here your task is it to calculate the background radiation temperature if all baryonic matter is converted into black body radiation.

3. On average the universe is very different from an ordinary office. Compute P_m/P_r in an office at $T = 20^o C$ and $P = 1$ bar.

4. In the box at the end of Chap. 9 we state that $d \ln T = -Y(x) d \ln a$, where $Y(x) = (1 + \frac{3}{4}x)/(1 + \frac{3}{8}x)$ and $x = N/(V\sigma T^3) = P_m/(3P_r)$. Derive this result.

5. This problem is a reminder of Bose- and Fermi-statistics from statistical mechanics. Its purpose is to make sure that the origin of the factors 7/8 in (9.34) and (9.35) or the factor 3/4 in (9.44) are reasonably clear.

   (a) Derive the occupation number for Bosons, i.e.

   $$n_k^{(B)} = \left(e^{\beta(\epsilon_k - \mu)} - 1\right)^{-1} ,$$

   where k is the index of the one-particle energy levels, ϵ_k is the attendant energy and μ is the chemical potential. $\beta = 1/T$, where we have set the Boltzmann constant equal to unity. Proceed as follows: (i) write down the number of distinct ways B_i to arrange N_i indistinguishable Bosons distributed over g_i distinguishable degenerate states. (ii) the equilibrium entropy is given by the maximum value of $S = \sum_i \ln B_i$. Assume that all factorials are large and can be approximated via Stirling's formula, i.e. $\ln N! \approx N \ln N - N$. Note that $n_i = N_i/g_i$ and express S in terms of n_i rather than N_i. (iii) find the equilibrium occupation number n_k by variation of $S + \lambda_1 \sum_i \epsilon_i N_i + \lambda_2 \sum_i N_i$. λ_1 and λ_2 are Lagrange parameters, which help to satisfy the conditions $E = \sum_i \epsilon_i N_i$ and $N = \sum_i N_i$, where E is the total energy and N is the total particle number. From thermodynamics we can see that $\lambda_1 = -\beta$ and $\lambda_1 = \beta\mu$.

   (b) Derive the occupation number for Fermions, i.e.

   $$n_k^{(F)} = \left(e^{\beta(\epsilon_k - \mu)} + 1\right)^{-1} ,$$

   following the exact same route. Note that B_i is now different and you should call the new combinatorial factor F_i.

   (c) Transform the summation over the one-particle energy states $\sum_i$ into an integral—first over the wave vector $\vec{k}$ and subsequently over ω via $\omega = ck$, where c is the speed of light. Write down the energy density E/V expressed as integral over ω for both Bosons and Fermions.

Chapter 10
Accelerated Expansion of the Universe

The most prominent effect of gravity in our current universe appears to be a uniform repulsion. In this chapter we shall introduce some of the basic concepts for a theoretical description of this astonishing discovery made at the end of the 20th century.

10.1 The Cosmological Redshift

We consider a ray of light coming towards us at the center of a Robertson-Walker coordinate system. The ray of light of course obeys $d\tau = 0$. According to (8.19) we have

$$dt = \pm a(t) \frac{dr}{\sqrt{1 - Kr^2}} . \tag{10.1}$$

The light comes towards us and thus dr is negative, i.e. we have to choose the minus sign. A photon emitted from a source at r_e at time t_e (index e for emitted) arrives at our position, i.e. $r_o = 0$, at time $t_o > t_e$ (index o for observed) given by[1]

$$\int_{t_e}^{t_o} \frac{dt}{a(t)} = \int_0^{r_e} \frac{dr}{\sqrt{1 - Kr^2}} = \begin{cases} \frac{\arcsin(\sqrt{K}r_e)}{\sqrt{K}} & K > 0 \\ r_e & K = 0 \\ \frac{\text{arcsinh}(\sqrt{|K|}r_e)}{\sqrt{|K|}} & K < 0 \end{cases} \tag{10.2}$$

The same equation holds when the lower limit t_e is replaced by $t_e + \delta t_e$ and the upper limit t_o is replaced by $t_o + \delta t_o$. There is no change on the right hand side of

---

[1] Note: $\sinh x = -i \sin(ix)$ and thus $\text{arcsinh}(\sqrt{|K|}r_e)/\sqrt{|K|} = \arcsin(\sqrt{K}r_e)/\sqrt{K}$ for $K < 0$.

© Springer Nature Switzerland AG 2020
R. Hentschke and C. Hölbling, *A Short Course in General Relativity and Cosmology*, Undergraduate Lecture Notes in Physics,
https://doi.org/10.1007/978-3-030-46384-7_10

the equation and thus[2]

$$\frac{\delta t_e}{a(t_e)} = \frac{\delta t_o}{a(t_o)}. \tag{10.3}$$

We can interpret $\nu_e \equiv 1/\delta t_e$ and $\nu_o \equiv 1/\delta t_o$ as being the frequencies of the departing and the arriving photon. This yields the equation

$$\boxed{1 + z = \frac{a(t_o)}{a(t_e)}}, \tag{10.4}$$

where $z = (\nu_e - \nu_o)/\nu_o$ is the frequency shift.

If the source is nearby, we may expand $a(t)$ at $t = t_o$, i.e. $a(t_e) = a(t_o) + \dot{a}(t_o)(t_e - t_o) + \ldots$, which yields

$$z = H(t_o)(t_o - t_e) + \cdots , \tag{10.5}$$

where H is the Hubble constant. Therefore we expect a redshift for $H(t_o) > 0$ or a blueshift for $H(t_o) < 0$ that increases linearly with the proper distance $d = t_o - t_e$ for galaxies close to us so that (10.5) holds. Roughly this means $z < 0.1$. But note that if z is too small, we do not measure the expansion (or contraction) of space but rather other types of velocities galaxies can have. In the following we are interested therefore in the large distances.

The first Friedmann equation in the form (9.26) can be rewritten as

$$dt = \frac{dx}{H_0 x \sqrt{\Omega_v + \frac{\Omega_K}{x^2} + \frac{\Omega_m}{x^3} + \frac{\Omega_r}{x^4}}} , \tag{10.6}$$

where $x = a/a_0 = 1/(1 + z)$. If we define the time origin as corresponding to an infinite redshift, then the time at which light was emitted that reaches us with redshift z is given by

$$\boxed{t(z) = \frac{1}{H_0} \int_0^{\frac{1}{1+z}} \frac{dx}{x \sqrt{\Omega_v + \frac{\Omega_K}{x^2} + \frac{\Omega_m}{x^3} + \frac{\Omega_r}{x^4}}}}. \tag{10.7}$$

Note that the age of the universe is $t(0)$, i.e. $z = 0$ (≈ 14 billion years). Thus, the quantities Ω as well as H_0 are of great significance.

---

[2]Note:

$$\int_{a+\delta a}^{b+\delta b} f(x)dx = F(b + \delta b) - F(a + \delta a) \approx F(b) - F(a) + f(b)\,\delta b - f(a)\,\delta a.$$

Inserting the Ω-values from Sect. 9.5 we find that the integral in (10.7) is very close to one, i.e. $t(0) \approx H_0^{-1}$. This means that the tangent at the point 'we are here' in the lower graph in Fig. 9.4 passes almost through the origin. If, on the other hand, we insert $\Omega_m = 1$, setting all other Ω equal to zero, then the integral yields about 0.67. This is too small, i.e. the age of the universe which we get using H_0 given in Appendix A is too small (about $9 \cdot 10^9$ y). But what this really shows is that Ω_v is necessary to obtain the proper age of the universe.

10.2 Accelerated Growth and Evidence for Dark Energy

There are different distance indicators (cf. the first chapter in [26]). Here we concentrate on only one—type Ia supernovae. This type of supernova is believed to occur when a white dwarf star accretes matter, pushing it beyond the Chandrasekhar limit, the maximum possible mass that can be supported by electron degeneracy pressure. The subsequent collapse and thermonuclear explosion are very well defined by the Chandrasekhar limit, which results in only little variation in the absolute luminosity of the explosion.

Let's talk about luminosity. The absolute luminosity L is the energy emitted per second by an object. In Euclidean geometry we define the apparent luminosity l via

$$l = \frac{L}{4\pi d^2} , \qquad (10.8)$$

where d is the distance to the object. At large distances this formula needs modification. (i) The energy of photons received is less than their energy when they were emitted by the redshift factor $a(t_e)/a(t_o) = 1/(1 + z)$; (ii) The rate of arrival compared to the rate of emission is reduced by the same factor (i.e. the space between photons is stretched as well); (iii) by the time of arrival, t_o, d is given by $r_e a(t_o)$, where r_e is the coordinate distance of the object as seen from the receiver. Thus (10.8) becomes

$$l = \frac{L}{4\pi d_L^2} , \qquad (10.9)$$

where

$$d_L^2 = a^2(t_o) r_e^2 (1 + z)^2 . \qquad (10.10)$$

We find r_e from (10.2), i.e.

$$a_o r_e(z) = \frac{\sinh\left[\sqrt{\Omega_K} a_o H_o \int_{t_e}^{t_o} \frac{dt}{a(t)}\right]}{H_o \sqrt{\Omega_K}} , \qquad (10.11)$$

where $a_o = a(t_o)$.[3] Thus, if we know L for just one type Ia supernova, then we know it for all others.[4] Then we measure l and z for as many type Ia supernovae as possible and plot l versus z. We can fit these data using (10.10) and (10.11). If we know how to express $a(t)$ in terms of the different densities of 'matter', we can adjust their relative fractions optimising the fit to the data. This in turn tells us whether the universe shrinks, grows and, if it grows, whether the growth accelerates.

In fact, we know how to express $a(t)$ in terms of the different densities of 'matter'. Based on (10.7) in differential form we obtain

$$a_0 H_0 \int_{t_e}^{t_o} \frac{dt}{a(t)} = \int_{\frac{1}{1+z}}^{1} \frac{dx}{x^2 \sqrt{\Omega_v + \frac{\Omega_K}{x^2} + \frac{\Omega_m}{x^3} + \frac{\Omega_r}{x^4}}} . \qquad (10.12)$$

Because this is a somewhat complicated formula, it is sensible to compare two special cases: (i) $\Omega_v = 1$, $\Omega_K = \Omega_m = \Omega_r = 0$ and (ii) $\Omega_K = 1$, $\Omega_v = \Omega_m = \Omega_r = 0$. In these cases the above integral is easy and using (10.10) we find

$$H_0 d_L(z) = \begin{cases} z + z^2 & (i): \quad \text{vacuum dominated} \\ z + \frac{z^2}{2} & (ii): \quad \text{empty} \end{cases} . \qquad (10.13)$$

The two curves are shown in Fig. 10.1. Here we have used logarithmic axes for the sake of similarity to the original data shown in Fig. 10.2, presenting initial evidence for dark energy, i.e. $\Omega_v > 0$. The overall conclusion based on this and other measurements is $\Omega_K = 0$, $\Omega_v \approx 0.7$, $\Omega_m \approx 0.3$, and $\Omega_r \approx 0$. The following figure, Fig. 10.3, shows the Supernova Cosmology Project data together with about 180 data points from a subsequent study. The lines are obtained using (10.10) to (10.12)—evaluating the integral in (10.12) numerically. The solid line appears to give the best fit. However, whether $\Omega_v = 0.7$ or 'merely' 0.6 or even 0.5 is difficult to decide based on our example with its limited number of data points.

Figure 10.4 shows the normalised scale factor, $a(t)/a_0$, versus normalised time, $H_0 t$, obtained by solving (10.7) numerically. The red dot indicates our current position (see also Fig. 9.4).

It is important to note that at this point we have all the equations and all the parameters (i.e. (7.6), (9.17), (9.21), and (9.40)), necessary to compute or check the numbers on the graphs such as shown in Fig. (7.1) or Fig. (9.4).

---

[3] Here we use (9.24) to express K in terms of Ω_K, a_0 and H_0 referring to today. The index o not always means 'today', even though most of the time it does. In the cases when it does not, (10.11) still holds—but with $\Omega_K = \Omega_{K,o}$.

[4] Actually, this is not quite correct. But the underlying problem can be fixed (see for instance [27]).

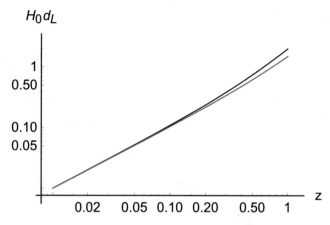

Fig. 10.1 The examples in (10.13). Black: (i) vacuum dominated; red: (ii) empty

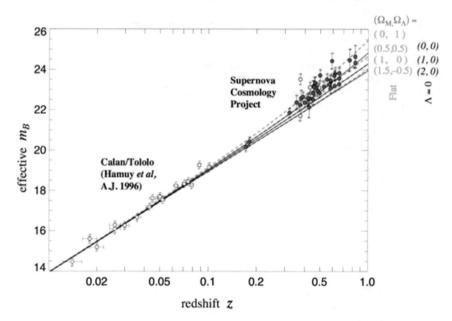

Fig. 10.2 Evidence for dark energy presented by S. Perlmutter et al. [28]. Here Ω_M is our Ω_m and Ω_Λ is our Ω_v. 'Effective m_B' is equal to a constant plus 5 log $H_0 d_L$. The different curves correspond the different models with $\Omega_K = 0$ and $\Omega_v = 0$

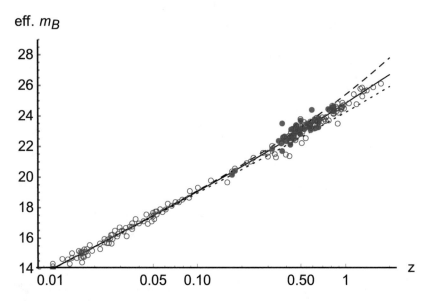

Fig. 10.3 The Supernova Cosmology Project data from the previous figure together with the data from Table 5 in [29]. The lines are obtained using (10.10) to (10.12). Solid Line: $\Omega_v = 0.7$, $\Omega_m = 0.3$, $\Omega_K = \Omega_r = 0$; dashed line: $\Omega_v = 1.0$ and $\Omega_m = \Omega_K = \Omega_r = 0$; dotted line: $\Omega_m = 1.0$ and $\Omega_v = \Omega_K = \Omega_r = 0$. Note that the Riess data as well as the theoretical curves are shifted vertically to match the Supernova Cosmology Project data

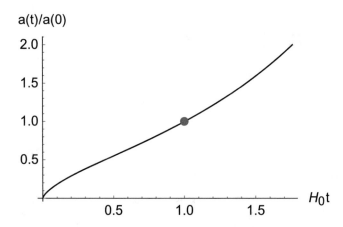

Fig. 10.4 Scale factor in reduced units versus time in units of $1/H_0$. The red dot indicates our current position. Here $\Omega_v = 0.74$ and $\Omega_m = 0.26$. Note that the cross-over from decelerating expansion of the universe to its accelerated expansion occurred roughly at a redshift of $z = 0.79$ corresponding to about half the current age of the universe (problem)

• **Example—Effective Repulsive Force between Galaxies due to the Presence of Dark Energy:** It is useful to get the feel of the magnitude of the effective repulsive force between galaxies due to the presence of dark energy. Here we consider the Milky Way and Andromeda. What is the ratio between the repulsive force due to dark energy, F_Λ, and the gravitational attraction between the two galaxies expressed via F_g?

First we compute their mutual gravitational attraction according to Newtonian gravity, which is an excellent approximation. Both Andromeda and the Milky Way have about equal mass $M = M_A \approx M_{MW} \approx 10^{12} M_\odot$ and are separated by a distance $r_A \approx 2.4 \cdot 10^{22}$ m. Thus

$$a_g = -G\frac{M}{r_A^2} \approx -23 \cdot 10^{-14} \text{ m/s}^2 \tag{10.14}$$

is the field strength of the gravitational attraction between the two.

To estimate the corresponding field strength resulting from the cosmological constant, we use (4.51) with $T_{\mu\nu} = 0$ which yields the Laplace equation

$$\Delta\varphi = -\Lambda . \tag{10.15}$$

The general solution can be straightforwardly obtained in cartesian coordinates as $\varphi = -\frac{1}{6}\Lambda r^2$ (cf. (4.39)), which corresponds to a repulsive force linear in r. In this case we find

$$a_\Lambda = -\partial_r\varphi|_{r_A} = \frac{\Lambda}{3}r_A = H_0^2\Omega_v r_A \approx 8.6 \cdot 10^{-14} \text{ m/s}^2 . \tag{10.16}$$

Therefore the ratio of the two respective forces is

$$\frac{F_\Lambda}{F_g} \approx -0.4 . \tag{10.17}$$

This estimate suggests that Andromeda and Milky Way will remain gravitationally bound (neglecting of course the current relative velocity of the two galaxies). However, it also means that galaxies significantly farther from us than Andromeda will eventually disappear, because in the future ρ_v dominates and thus

$$\frac{\dot{a}}{a} = \sqrt{\frac{8\pi G}{3}\rho_v} = H . \tag{10.18}$$

Hence, according to (9.7),

$$D_h = \frac{1}{\sqrt{\frac{8\pi G}{3}\rho_v}} \ . \tag{10.19}$$

Everything, except nearby and gravitationally bound galaxies, will disappear behind this event horizon. Finally it is important to note that similar comparisons can be made on the scale of the solar system. Is, for instance, the perihelion precession of Mercury affected by dark energy? The answer is that the effects of dark energy are entirely insignificant on the scale of the solar system (see also Problem 2 in Appendix D).

Before we move on to the next section let us provide a precise equation for the above quantity D_h. Our starting point is (10.1), i.e.

$$\frac{dt}{a(t)} = \pm \frac{dr}{\sqrt{1 - Kr^2}} \ . \tag{10.20}$$

The comoving or coordinate distance traveled by a photon emitted at time t_e and observed at time t_o is given by

$$l(t_o, t_e) \equiv \int_{t_e}^{t_o} \frac{dt}{a(t)} \ . \tag{10.21}$$

The earliest possible emission time is $t_e = 0$ and the attendant (physical) horizon distance of the observer is

$$D_{h,p}(t_o) = a(t_o)l(t_o, t_e = 0) \ . \tag{10.22}$$

This is the maximum distance at which past events can be observed at t_o.

But there is also a another case. This is the maximum distance, also called event horizon, which the cosmologist existing at t_e can 'observe' in the future when $t_o \to \infty$. It is given by

$$D_{h,f}(t_e) = a(t_e)l(t_o = \infty, t_e) \ . \tag{10.23}$$

• **Example—Event Horizon Distance in a Dark Energy Dominated Universe:** We want to compute $l(t_o = \infty)$, $t_e =$ today) assuming that $\rho = \rho_o$, i.e. vacuum energy dominates. According to the first Friedmann equation

$$a(t) = a(t_e) \exp[H(t - t_e)] \ , \tag{10.24}$$

where $H = \sqrt{(8\pi G/3)\rho_o}$. Inserting this into (10.21) we obtain

$$l(\infty, t_e) = \frac{1}{a(t_e)} \int_{t_e}^{\infty} dt \, \exp[-H(t - t_e)] = \frac{1}{a(t_e)} \frac{1}{H} . \qquad (10.25)$$

The attendant physical distance is

$$D_{h,f}(t_e) = a(t_e)l(\infty, t_e) = \frac{1}{H} , \qquad (10.26)$$

which confirms our previous result (10.19), but gives a precise meaning to D_h.

10.3 Structure in the Cosmic Microwave Background

We have discussed in Sect. 9.1 that the CMB played an important role in the validation of the Big Bang model. The fact that it is an almost ideal black body with a temperature of 2.725 K is strong evidence in support of a hot beginning of our expanding universe. But there is much more information to be extracted from the tiny temperature fluctuations and its polarization, which have been measured with ever increasing precision. We will give here a short and for the most part qualitative overview of the main features of the CMB.

10.3.1 Correlations of Temperature Fluctuations

Figure 10.5 is a map of $\delta T(\theta, \phi)$, i.e. the local deviation of the temperature from its average over the entire sky in the direction defined by the angles θ and ϕ. Even more interesting is the product $\delta T(\theta, \phi)\delta T(\theta', \phi')$. The average of this product over the entire sky, i.e.

$$C(\gamma) = \langle \delta T(\theta, \phi)\delta T(\theta', \phi') \rangle , \qquad (10.27)$$

is a temperature fluctuation correlation function. But what is the meaning of γ? The quantity γ is the angle defined by the two unit vectors $\vec{e}(\theta, \phi)$ and $\vec{e}\,'(\theta', \phi')$. This assumes that the above average product depends solely on the angular separation between the points in the sky at which we measure the temperature and not on the choice of axes of our coordinate system.

What is the physical meaning of $C(\gamma)$? If γ is really small, then we expect that $\delta T(\theta, \phi) \approx \delta T(\theta', \phi')$. Thus, $C(\gamma)$ has a certain positive value in this limit. If however γ is large, then we expect that $\delta T(\theta, \phi)$ and $\delta T(\theta', \phi')$ are not related at all. In

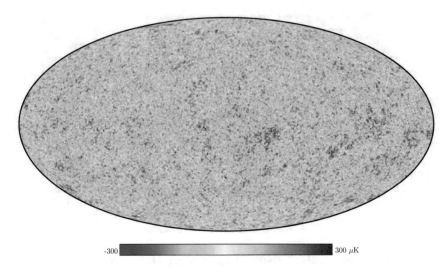

-300   300 μK

Fig. 10.5 Map of the temperature fluctuations in the cosmic microwave background (CMB). The average temperature is 2.725 K. The deviations from this average, here indicated by the different colours, are in the 10^{-4} K range. Image: ESA and the Planck collaboration [30]

fact, because temperature deviations can be negative and positive, we expect $C(\gamma)$ to vanish. A simple $C(\gamma)$ satisfying these conditions looks as follows:

$$C(\gamma) = A \exp[-\gamma/\gamma_o] , \qquad (10.28)$$

where A is a constant. This $C(\gamma)$ rapidly approaches zero when $\gamma > \gamma_o$. We might call γ_o a fluctuation correlation angle. Points in the sky whose angular separation exceeds γ_o quickly become uncorrelated. An alternative $C(\gamma)$ might have the power-law form

$$C(\gamma) = \frac{B}{\gamma^p} , \qquad (10.29)$$

where B and p (> 0) are constants. The angular range of this $C(\gamma)$ is 'infinite'. Note that (10.28) describes fluctuation patterns characterised by a distinct angular spread γ_o, whereas (10.29) describes fluctuation patterns possessing no particular scale.[5]

We can expand $C(\gamma)$ in terms of Legendre polynomials, i.e.

---

[5]This is quite crude. For instance, we do not expect $C(\gamma)$ to diverge in the limit $\gamma \to 0$ as in the case of the power law. The opposite limit is also problematic, because we are dealing with a finite geometry. We also like to add the following caveat. In statistical mechanics fluctuation correlation functions at critical points, where the fluctuation correlation length is infinite, are usually described by power laws. In the continuum limit a system at criticality is described as scale invariant. So in this context power laws are quite generally associated with scale invariance. In cosmology this term is special and we return to scale invariance in Chap. 11. The general point, i.e. scale invariance in statistical mechanics versus scale invariance in cosmology, is discussed in [31].

$$C(\gamma) = \sum_{l=0}^{\infty} \frac{2l+1}{4\pi} C_l P_l(\cos\gamma) \ , \tag{10.30}$$

where

$$\frac{1}{4\pi} C_l = \frac{1}{2} \int_0^\pi d\gamma \sin\gamma \, C(\gamma) P_l(\cos\gamma) \tag{10.31}$$

(see for instance [9]; Sect. 3.3). The expansion coefficients C_l contain all the physical information.[6] It is customary to present this information expressed in functions like

$$\Delta_T = \left(\frac{l(l+1)}{2\pi} C_l \right)^{1/2} . \tag{10.32}$$

The upper panel of Fig. 10.6 shows Δ_T versus l for the mock fluctuation correlation function (10.28) using three different γ_o. Note that γ_o and the value of l at which Δ_T has a maximum, l_{max}, satisfy

$$\gamma_o \approx \frac{\pi/2}{l_{max}} . \tag{10.33}$$

The lower panel of Fig. 10.6 also shows Δ_T versus l for the exponential, the power-law and a combined fluctuation correlation function. Note that the pure power-law forms yield an almost constant Δ_T in the range of l-values shown here.

Δ_T based on the analysis of the cosmic background temperature fluctuation map (cf. Fig. 10.5) is depicted in Fig. 10.7. These data of course are more complicated and the underlying correlations are far from simple. Nevertheless, it is possible to compute the different contributions to the overall $C(\gamma)$, providing much support for our standard model of cosmology including inflation. How this can be done in principle is discussed in the next chapter.

Remark The scatter in Fig. (10.7) is large when l is small and small when l is large. What is the reason for this? We just saw that there is an inverse relation between the patch size defined by a certain γ and the attendant l. Small l correspond to large patches in the sky, whereas large l correspond to small patches in the sky. However, there are many more independent small patches than there are large patches!

---

[6]Note that the $l = 0$ term is a constant, which can and should be eliminated by a proper definition of the average temperature. The $l = 1$ or dipole term is mainly determined by the Doppler shift due to our motion through space and therefore subtracted in the present context.

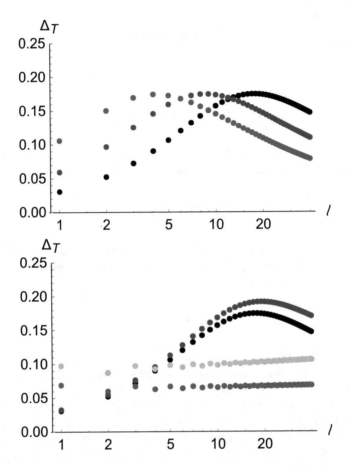

Fig. 10.6 Top: Δ_T versus l for the mock fluctuation correlation function (10.28). Here $A = 1$ and $\gamma_o = \pi/40$ (black), $\pi/20$ (blue), $\pi/10$ (red). Bottom: Δ_T versus l. Black: $C(\gamma) = \exp[-\gamma/\gamma_o]$ with $\gamma_o = \pi/40$; blue: $C(\gamma) = \gamma^{-p} \exp[-\gamma/\gamma_o]$ with $\gamma_o = \pi/40$ and $p = 0.05$; red: $C(\gamma) = \gamma^{-p}$ with $p = 0.05$; green: $C(\gamma) = \gamma^{-p}$ with $p = 0.1$

10.3.2 The Sachs-Wolfe Effect

The temperature fluctuations in the CMB are related to density fluctuations and the fluctuations in the gravitational potential. But what is the mathematical form of this relation? Let us start with the simplest case and assume small fluctuations, which did not dynamically evolve but expanded adiabatically. We have denser regions with a higher gravitational potential and less dense region with a lower potential. It will cost photons energy to climb out of the potential well of the overdense regions. If the

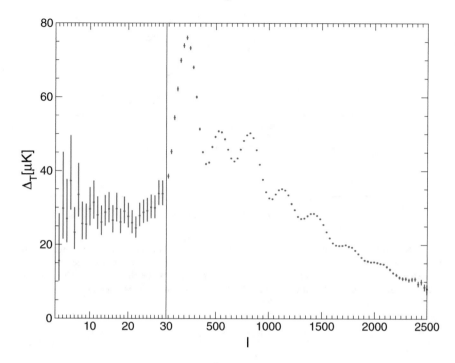

Fig. 10.7 Δ_T versus l based on the analysis of the cosmic background temperature fluctuation map as observed by the Planck satellite. Note that modes $l > 30$ are binned for better visibility and the scale of the x-axis differs for them. Data courtesy of the Planck Public Data Release 2 (https://irsa.ipac.caltech.edu/data/Planck/release_2/ancillary-data/). The results are presented in [18, 30]

potential difference is $\delta\phi$,[7] then they will be redshifted by a factor $\delta\nu/\nu = \delta\phi$ (cf. (2.83)). But there is a second effect we need to consider. Inside an overdense region with a deeper potential, time is slowed down by the same factor, i.e. $\delta t/t = \delta\phi$. The cooling of the hot cosmic medium will therefore be slightly delayed in an overdense bubble and decoupling will occur at a larger $a(t)$. At the time of decoupling the universe is matter dominated, i.e. $a \propto t^{2/3}$. Therefore $\delta a/a = (2/3)\delta t/t$ and thus the redshift of the arriving photon will be reduced by a factor $\delta\nu/\nu = -(2/3)\delta\phi$. Here we use (10.4), i.e. $-\delta\nu/\nu = \delta a/a$. Combining the two effects we obtain a redshift from an overdense region

$$\frac{\delta\nu}{\nu} = \frac{1}{3}\delta\phi \ . \tag{10.34}$$

For thermal radiation photons a redshift implies a reduced radiation temperature and $\delta T/T = -\delta a/a = \delta\nu/\nu$. Thus

---

[7]Note that $\delta\phi$ is the deviation from the average potential. Overdensity implies $\delta\phi < 0$, as we can see from the Poisson equation in Fourier space, i.e. $-k^2\delta\phi_k \sim \delta\rho_k$.

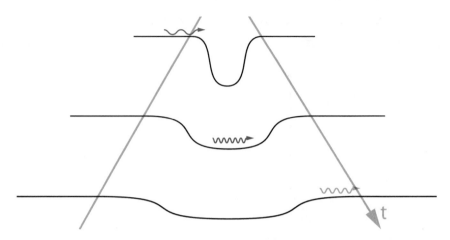

Fig. 10.8 Illustration of the integrated Sachs-Wolfe effect. A photon enters a steep potential well (top) and is blueshifted (middle). When exiting the well, cosmic expansion in the presence of vacuum energy has made the well shallower and the blueshift is not fully undone. Note that the usual redshift $\lambda \propto a(t)$ of a homogenous universe has been ignored for simplicity

$$\boxed{\frac{\delta T}{T} = \frac{1}{3}\delta\phi}\,, \tag{10.35}$$

which is known as the Sachs-Wolfe effect. A region of reduced temperature in the CMB can thus be directly identified as an overdense region with reduced gravitational potential at decoupling.

Once the photons leave the surface of last scattering they will also pass through regions of different gravitational potential before they reach our antennae. These will cause temperature shifts that need to be integrated along the entire path termed the integrated Sachs-Wolfe (ISW) effect. At first sight one might think that the energy of a photon is oblivious to the potential landscape which it traverses during its journey—because after all these are potentials. Every hill it climbs it will later come down from and every valley it enters it will have to come out of eventually—but this is not as simple in an expanding universe with vacuum energy. Let us imagine that our photon enters a large scale potential well, e.g. a supercluster of galaxies (Fig. 10.8). Upon entering the well it gains energy and gets blueshifted relative to a photon that avoids falling into the well. If the photon remains within the well for a sufficiently long time, the cosmic expansion in the presence of vacuum energy can make the well shallower and broader. Therefore it will have to spend comparatively less energy to finally climb out of the well. This results in a remnant blueshift compared to a photon that did not fall into the well. As the blueshift depends on the vacuum energy, this so called late time ISW in principle offers an independent way of determining the vacuum energy density at the time the photon crossed the potential well.

Of course, the resulting temperature fluctuation is superimposed on top of the temperature fluctuation at the time of decoupling (and all sorts of astrophysical disturbances) and therefore is not easy to isolate. It is however possible to pick out the component of the temperature fluctuation in the CMB that correlates with the known matter distribution from galaxy counts. In this way direct evidence for the existence of vacuum energy has been obtained albeit with rather small statistical significance.

10.3.3 The Size of Visible Structures

Let's discuss the observation of spots or patches in the sky. Figure 10.9 depicts an astronomer observing a spot of proper diameter δl perpendicular to his line of sight. The equation

$$d_A = \frac{\delta l}{\delta \gamma} \tag{10.36}$$

defines the angular-diameter distance. To find out how to calculate d_A theoretically we study the space-part of the FRW metric, i.e.

$$ds = a(t_e) r_e d\theta . \tag{10.37}$$

The meaning of t_e and r_e is the same as in the context of (10.1). In particular t_e is the time at which the spot size is measured at its position. Now we can basically equate ds with δl and $d\theta$ with $\delta \gamma$. Thus

$$\delta l = a(t_e) r_e \delta \gamma = \underbrace{\frac{a(t_o)}{1+z} r_e}_{=d_A} \delta \gamma . \tag{10.38}$$

And using (10.10) we find

$$d_L = d_A (1+z)^2 . \tag{10.39}$$

From (10.2) and (10.22) we obtain in the limit $\Omega_K \to 0$ and for large z

$$d_L \approx z D_{h,p} \tag{10.40}$$

or

$$D_{h,p}(t_o) \approx z d_A . \tag{10.41}$$

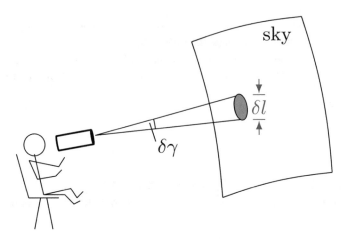

Fig. 10.9 An astronomer at the origin observes a 'spot' of proper diameter δl perpendicular to his line of sight

This relations combines the horizon distance D_{hp} at the time of observation,[8] the redshift of the observed photons and the attendant d_A.

Now we are ready to try out some numbers. According to (9.13) the 'last scattering' occurred $4 \cdot 10^5$ years after the Big Bang. Light we observe coming from the surface of last scattering therefore has a redshift

$$z_{ls} \approx 1040 . \tag{10.43}$$

This number is based on (10.7) and our Benchmark Model ($\Omega_v = 0.7$, $\Omega_m = 0.3$, $\Omega_r = 9 \cdot 10^{-5}$ and $\Omega_K = 0$). Next we compute the horizon distance today and we find

$$D_{h,p}(t_0) \approx 13900 \text{ Mpc} . \tag{10.44}$$

Inserting z_{ls} and $D_{h,p}(t_0)$ into (10.41) we obtain d_A for our astronomer, i.e.

$$d_A \approx 13.4 \text{ Mpc} . \tag{10.45}$$

But what is a reasonable spot size δl? Let's suppose we concentrate on the largest peak in Fig. 10.7, corresponding to the lowest l-value (as compared to the other peaks) and therefore to the largest angular correlation distance (aside from the underlying

---

[8]Which we evaluate via

$$D_{h,p}(t) = \frac{1}{H_o(1 + z(t))} \int_0^{1/(1+z(t))} \frac{dx}{x^2 \sqrt{\Omega_v + \frac{\Omega_K}{x^2} + \frac{\Omega_m}{x^3} + \frac{\Omega_r}{x^4}}} \tag{10.42}$$

(Problem 3).

(almost) scale invariant correlations which we shall discuss in the context of inflation). The structures we observe in the CMB have their origin at the Big Bang and
thus the horizon distance at the time of last scattering, i.e.

$$D_{h,p}(t_{ls}) \approx 0.27 \text{ Mpc} \tag{10.46}$$

is a measure for the largest possible correlation distance. To see this we return to Fig.
9.3. Here $D_{h,p}(t_{ls})$ corresponds to the projection of the small arrow, representing a
photon emitted by B at the Big Bang, onto the green circle or the distance from B to
A along this circle (neglecting expansion!). Had the event taking place in B at the Big
Bang sent photons in every direction, then there would be another photon detected at
A' on the other side of B along the green circle. In other words, the photons seen at A
and A' are correlated and their correlation distance is $2D_{h,p}(t_{ls})$. But what if the event
at B did not produce photons propagating with speed c but a signal propagating at a
slower speed c_s. In this case the correlation distance is smaller, i.e. $2D_{h,p}(t_{ls})(c_s/c)$.
Therefore we must figure out the physical nature of the event which took place in B
following the Big Bang.

10.3.4   Baryonic Acoustic Oscillations

Shortly after the Big Bang the universe contained a plasma of photons, electrons
and light nuclei. The photons scattered strongly off the electrons and the electrons
coupled strongly to the nuclei due to their charge. All in all there existed a well
equilibrated fluid (or plasma) of ionised matter and photons. This is exactly what we
had studied in the second box in Sect. 9.6. The total pressure in this fluid was almost
entirely due to the radiation pressure, i.e. $P = P_r$. The event at B was the emission
of a sound wave in this fluid.

This is a good place to mention two important time scales. Simple dimensional
analysis tells us that the collapse time t_c of a self-gravitating system or region should
be given by

$$t_c \sim (G\rho_m)^{-1/2} . \tag{10.47}$$

Note that here ρ_m is the (average) mass density in the system. If the collapse is to be
prevented by building up an attendant pressure gradient, then the necessary time for
this, t_{pres}, should satisfy $t_c > t_{pres}$. The time t_{pres} is determined by

$$t_{pres} \sim R/c_s , \tag{10.48}$$

where R is the radius of the region and c_s is the speed of sound. The particular
R_J for which $t_c = t_{pres}$ is called the Jeans length. Overdense regions larger than R_J
collapse, whereas overdense regions smaller than R_J merely oscillate. The latter is
caused by the pressure buildup during the collapse, which slows it down and then

leads to another expansion. This produces baryonic acoustic oscillations (BAO). And even though dark matter is an important ingredient to the sound wave generation via its gravity, it did not take part in what we talk about next, due to its weak coupling to ordinary matter as well as to radiation.

Let's assume that we look at a region with overdensity from the very start. Its rapid expansion, essentially with the speed of sound in the photon-electron-baryon plasma, c_s, produces a shell with increased baryon density travelling outward from B. Note that c_s in this fluid is given by $c_s \approx 1/\sqrt{3}$ (in units of the speed of light!) as we show in a box at the end of this section. So the spherical surface defined by the density crest continues to grow until after $4 \cdot 10^5$ years the temperature in the universe allows recombination to occur. The baryons, which thus far were dragged along by the photons, suddenly decouple from the photons. By themselves they cannot maintain the above c_s. In fact, they stop at this point, because their own pressure, as we have shown, is negligible if compared to the radiation pressure, i.e. $c_{s,b} \approx 0$. Thus, there is a huge bubble with overdense matter on its surface. And because there are many spots B, where such bubbles emanate, they end up forming a network of interpenetrating bubbles arrested in space at the time of recombination, which may still be observed today. After recombination the dark matter distribution, which originally is thought to have been concentrated in the center of the overdense region, from which the sound wave emanated, and the baryonic mass distribution, mainly located in the walls of the bubbles, began to coalesce. Even though this affected the baryon density in the walls of the aforementioned bubbles, it did not prevent the development of the baryon overdensity into large scale structures manifesting themselves today in the distribution of galaxies.

Since the time of recombination the universe has expanded by a factor $\sim z_{ls} \approx 1040$ (cf. (10.43)) and thus, using (10.46) and $c_s \approx 1/\sqrt{3}$, we expect these current structures in the distribution of galaxies to have an approximate size of

$$1040 \frac{0.27}{\sqrt{3}} \text{ Mpc} \approx 160 \text{ Mpc} . \tag{10.49}$$

Structural correlations of this size were first found in 2005 by the Two-degree-Field Galaxy Redshift Survey (2dF) and the Sloan Digital Sky Survey (SDSS) and have since been corroborated by other similar three dimensional surveys of our cosmic neighbourhood. Figure 10.10 shows the correlation function of galaxy density as measured by the Baryon Oscillation Spectroscopic Survey (BOSS) collaboration. The peak due to the above mechanism is clearly visible. Note that the correlation function is plotted versus comoving distance rather than physical distance because the structures have linear sizes $\propto 1 + z$ and the survey covers a region out to $z \sim 0.7$. This might look like a minor nuisance, but it is actually a great opportunity. The size of the BAO structures provides a standard ruler and if one can trace them throughout the evolution of the universe, the expansion history can be inferred from it.

Before we explore the baryonic acoustic oscillations more closely let us briefly return to the spot size. The size of the aforementioned bubbles is roughly the sound wave horizon distance at the time of last scattering. The angular width of this 'spot'

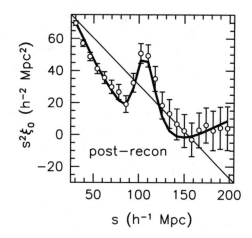

Fig. 10.10 Correlation function of galaxy density versus comoving distance in a sample of nearly one million galaxies in a volume of 13Gpc^3. A peak in the correlation function at the expected distance scale ≈ 160 Mpc is clearly visible. Note that $h^{-1} = 100\,\mathrm{km\,s}^{-1}$ Mpc$^{-1}/H_0$. Adapted from [32]

is given by

$$\delta\gamma \approx \frac{0.27/\sqrt{3}\ \mathrm{Mpc}}{13.4\ \mathrm{Mpc}} \approx 0.012\ \mathrm{rad}\ . \tag{10.50}$$

The next step in our chain of reasoning is the relation between $\delta\gamma$ and (10.33) obtained for our mock correlation function. Setting $\gamma = \gamma_o$ means that the correlation function is $C(\gamma = \gamma_o)/C(0) \approx 0.4$, i.e. it has decayed to about half of its maximum value. Increasing γ to $n\gamma_o$ yields $C(2\gamma_o)/C(0) \approx 0.14$ or $C(3\gamma_o)/C(0) \approx 0.05$. Thus $\delta\gamma \approx 2\gamma_o$ appears to be a sensible estimate for the largest angular range of correlations. Applying this to (10.33) yields

$$l_{max} \approx \pi/\delta\gamma \approx 260\ . \tag{10.51}$$

This is in accord with the position of the largest peak in Fig. 10.7.

Remark 1 The first peak in Fig. 10.7 roughly marks the boundary between large-scale inhomogeneities that have not changed since inflation, which we discuss in the next chapter, and the small-scale perturbations that have undergone substantial modification by gravitational instability.

Remark 2 It is also worth noting that because this result is obtained for $\Omega_K = 0$, it is an indication of the large scale flatness of the universe.

- **Velocity of Sound in Gases and Liquids:** A sound wave causes small perturbations in the pressure and the mass density, i.e. $P(\vec{r}, t) = P + \delta P(\vec{r}, t)$ and $\rho(\vec{r}, t) = \rho + \delta\rho(\vec{r}, t)$. Here P and ρ are position and time independent averages. We assume that these fluctuations are fast in the sense that they do not involve heat transfer (adiabatic).

The continuity equation, $\dot{\rho}(\vec{r}, t) + \vec{\nabla}\left[\rho(\vec{r}, t)\vec{u}(\vec{r}, t)\right] = 0$, yields

$$\delta\dot{\rho}(\vec{r}, t) + \rho\vec{\nabla}\vec{u}(\vec{r}, t) \approx 0 \quad \text{or} \quad \delta\ddot{\rho}(\vec{r}, t) + \rho\vec{\nabla}\dot{\vec{u}}(\vec{r}, t) \approx 0\,,$$

where $\vec{u}(\vec{r}, t)$ is the velocity of a mass element in the medium. Note that $u(\vec{r}, t)$ also is a small quantity. Here and in the following we keep the leading order contributions only. From the equation of motion of a volume element dV, $\rho(\vec{r}, t)dV\dot{\vec{u}}(\vec{r}, t) \approx -\vec{\nabla}P(\vec{r}, t)dV$, we find

$$\delta\ddot{\rho}(\vec{r}, t) - \vec{\nabla}^2\delta P(\vec{r}, t) \approx 0\,.$$

Using the isentropic compressibility $\kappa_S = -(1/V)(\partial V/\partial P)_S$, i.e.

$$\kappa_S = -\rho\left(\frac{\partial}{\partial P}\rho^{-1}\right)_S \approx \frac{1}{\rho}\frac{\delta\rho(\vec{r}, t)}{\delta P(\vec{r}, t)}\,,$$

we obtain the result

$$\delta\ddot{\rho}(\vec{r}, t) - \frac{1}{\kappa_S\rho}\vec{\nabla}^2\delta\rho(\vec{r}, t) \approx 0\,. \tag{10.52}$$

Inserting the plain wave solution $\delta\rho(\vec{r}, t) = \delta\rho_o\sin(\omega t - \vec{k}\cdot\vec{r})$ yields the velocity of sound $c_s \equiv \omega/k$:

$$c_s = \frac{1}{\sqrt{\kappa_S\rho}} \approx \sqrt{\delta P/\delta\rho}\,. \tag{10.53}$$

Applying this to the case at hand, we replace the mass density ρ by the total energy density of the baryon-photon plasma, $\rho_b + \rho_\gamma$, where as usual we use $c = 1$. Therefore the velocity of sound in the plasma is

$$c_s \approx \sqrt{\delta P/\delta(\rho_b + \rho_\gamma)} = \sqrt{\frac{\frac{4}{3}\sigma_\gamma T^3\delta T}{\delta\rho_b + 4\sigma_\gamma T^3\delta T}}\,.$$

Because $\rho_b \propto a^{-3} \propto T^3$, we have $\delta\rho_b = 3\rho_b\delta T/T$ and thus

$$c_s \approx \frac{1}{\sqrt{3(1 + (3/4)\rho_b/\rho_\gamma)}}\,. \tag{10.54}$$

Note that the ratio ρ_b/ρ_γ does depend on time. It is much smaller than one right after inflation, but at the surface of last scattering it is already close to one (Problem 2). Here we are interested in a rough estimate only and thus we use $c_s \approx 1/\sqrt{3}$.

We now have a qualitative understanding of a contribution to the origin of the first and largest peak in the power spectrum of CMB anisotropies. But Fig. 10.7 contains additional peaks whose amplitude seems to decrease rapidly at higher l. Where do these features come from?

Thus far we have not really discussed acoustic oscillations. However, in the context of (10.47) and (10.48) we have mentioned a mechanism, giving rise to density fluctuations localised to certain regions in space 'oscillating' between maximum compression and maximum rarefaction. In the next section we shall discuss that the density fluctuations right after the Big Bang existed on all length scales. But 'fluctuations on all length scales' is not what we expect to produce the distinct peaks in Fig. 10.7. In order to understand why there are distinct peaks, we assume, for simplicity, that the frequency of oscillation are described by (10.48),

$$\omega(R) \sim c_s / R \ . \tag{10.55}$$

This allows for an almost arbitrary range of frequencies $\omega(R)$ and attendant 'oscillating patch' sizes R. However, oscillators spend most of their time near their turning points. If an oscillator stops oscillating at an arbitrary time t_{LS}, it is likely close to maximum compression or maximum rarefaction in the present context. The time t_{ls} when when all oscillations come to a halt, is the time of decoupling of the photons or recombination. Thus, if we describe the oscillations via $\cos(\omega(R)t)$, then $\omega(R)t_{ls} \approx n\pi$, where $n = 1, 2, \ldots$. This is satisfied by distinct

$$R_n \sim n^{-1} \tag{10.56}$$

only, because c_s is the same for all frequencies.[9] In particular we expect $R_n/R_{n+1} \approx (n+1)/n$. And because $R_n/R_{n+1} = l_{n+1}/l_n$, where l_n is the position of the nth peak in Fig. 10.7, we expect $l_2/l_1 \approx 2$, $l_3/l_2 \approx 1.5$, $l_4/l_3 \approx 1.33$, $l_5/l_4 \approx 1.25$, $l_6/l_5 \approx 1.20$, $\ldots$. Using the peak positions from Fig. 10.7, i.e. $l_1 \approx 220, l_2 \approx 535, l_3 \approx 825, l_4 \approx 1110, l_5 \approx 1430$, and $l_6 \approx 1730$, we find $l_2/l_1 \approx 2.43$, $l_3/l_2 \approx 1.54$, $l_4/l_3 \approx 1.35$, $l_5/l_4 \approx 1.29$, and $l_6/l_5 \approx 1.21$. This is of course rough but not bad at all. In addition we note that the even-numbered peaks in Fig. 10.7 are less pronounced

---

[9]By choosing the cos solution we have tacitly implied an initial condition for the phases of all oscillators, namely that they start from their respective maxima. This is a physically sensible starting condition for an extremely hot plasma where radiation is dominant after inflation. We will see in the next chapter that inflation basically blows up quantum fluctuations of density by a large factor without providing significant relative momentum that would be characteristic of a sin mode. We also make the reasonable assumption (known as adiabatic) that density fluctuations for all species align. In principle the overdensity of one species could also initially be compensated by an underdensity of another. This would result in no curvature perturbation initially and thus the modes are called isocurvature. The latest Planck results constrain their contribution to below $\sim 2\%$.

in comparison to their odd-numbered neighbors. Can we understand this at least qualitatively?

Qualitatively we can describe the oscillations in terms of a simple harmonic oscillator equation

$$\ddot{\theta} + \frac{k^2}{3}(\theta + \psi) = 0 , \tag{10.57}$$

where $\theta \equiv \delta T/T$, i.e δT is a temperature fluctuation relative to some average temperature T. Note that this is what we do get from (10.52) if we describe $\delta\rho(\vec{r}, t)$ in terms of Fourier modes $\rho_k(t)e^{i\vec{r}\cdot\vec{k}}$, which means replacing $\vec{\nabla}^2\delta\rho(\vec{r}, t)$ by $-k^2\rho_k(t)$.[10] In addition we use (10.54), i.e. $c_s^2 = (3(1 + Q))^{-1}$, with $Q = (3/4)\rho_b/\rho_\gamma$. However, momentarily we use $Q = 0$, i.e. we do not consider the baryons at all. Nevertheless, the (assumed) adiabatic compression-expansion-compression density oscillations of the plasma translate into temperature fluctuations described by θ. The quantity ψ is a constant gravitational potential, which shifts the zero of the oscillations—akin to the shift of the equilibrium position of a classical harmonic oscillator suspended in a constant gravitational field on Earth. We can tie this k-space representation to our above real space considerations via $k \sim 1/R$. Just as before, we also use the solution $\theta + \psi = \frac{1}{3}\psi\cos(\omega(R)t_{ls})$ at last scattering where $\omega(R)t_{ls} = \pi n$ $(n = 1, 2, \dots)$. The amplitude, $\frac{1}{3}\psi$, follows according to the Sachs-Wolfe relation (10.35). Thus far, the magnitude of the oscillation amplitude is the same for all values of n. This changes when we add in the baryons, i.e. the baryon drag, which modifies (10.57) to

$$\ddot{\theta} + \frac{k^2}{3(1 + Q)}(\theta + (1 + Q)\psi) = 0 . \tag{10.58}$$

The perhaps unexpected factor $(1 + Q)$ multiplying ψ accounts for the fact that the gravity experienced by the photons, which carry the information regarding the temperature fluctuations, remains essentially unaffected. Now the solution becomes

$$\theta + \psi = -Q\psi + \frac{1}{3}\psi(1 + 3Q)\cos(\omega'(R)t_{ls}) . \tag{10.59}$$

The prime is a reminder that here the velocity of sound is different. Inserting again $\omega'(R)t_{ls} = \pi n$ $(n = 1, 2, \dots)$, we find

$$\theta + \psi\Big|_{\text{peak}} = \begin{cases} -\frac{1}{3}\psi\,(1 + 6Q)\,\cos(..) = -1 \\ \frac{1}{3}\psi \qquad\qquad \cos(..) = +1 \end{cases} . \tag{10.60}$$

At least crudely this explains the odd-even effect observed for the peak heights in Fig. 10.7. It also illustrates how this can be possibly used to infer information about

---

[10]The time t should really be conformal time.

the baryon content of the universe from the comparison of peak heights (especially comparing the second to the third peak).[11]

Finally let us discuss the decrease at high l. Remember that these are the extreme short wavelength modes, so they represent temperature differences at relatively small scales. But the photons are not completely stationary within the plasma—they diffuse. In addition, recombination does not happen instantaneously but takes a certain amount of time. During this time especially, photons perform a random walk and diffuse. Fluctuation on length scales below this diffusion length are thus suppressed and lead to the damping in the tail. Our discussion of the baryonic acoustic oscillation is quite rough and omits a number of important aspects. A well written article, which we recommend in this context, is [33].

The acoustic peaks are remarkable structures in the CMB spectrum, but there is something possibly more remarkable. At low $l \lesssim 30$, the fluctuations are approximately constant but not zero.[12] They represent correlations between points which, according to the classical Big Bang model , have never been in causal contact. This is another hint at an earlier, inflationary epoch of our universe. But there is more. Since these modes have never been in causal contact in the classical Big Bang model, it follows that whatever correlations exist between them must originate from the epoch that preceded it, e.g. from inflation. In that respect the flatness of Δ_T versus l at low $l \lesssim 30$ is a clear hint that initial fluctuations, before the universe entered the plasma phase, were (almost) scale invariant. This, in fact, was predicted by inflation before the CMB anisotropies were observed.

We have thus traced back the origin of structures in our universe to (almost) scale invariant inhomogeneities in a well equilibrated, almost homogenous and isotropic plasma in an almost flat universe. At this point we leave the realm of firmly established theories and turn to what many see as the currently most promising candidate theory to explain these peculiar initial conditions: inflation.

10.4 Problems

1. (a) Calculate the age of the universe based on (10.7) and $H(t_0) = 70$ kms^{-1}Mpc^{-1} for model (i) with $\Omega_v = 0.7$, $\Omega_K = 0$, $\Omega_m = 0.3$, and $\Omega_r = 9 \cdot 10^{-5}$. Repeat this calculation for model (ii) with $\Omega_v = 0.74$, $\Omega_K = 0$, $\Omega_m = 0.26$, and $\Omega_r = 9 \cdot 10^{-5}$.
   (b) Calculate the cross-over time t_{rm} from radiation to matter dominance defined by $\rho_m(t_{rm})/\rho_r(t_{rm}) = 1$. Provide t_{rm} for both models in part (a).

---

[11] The most recent numbers from the PLANCK satellite imply $\Omega_b \simeq 0.045$.

[12] At very small $l \lesssim 5$ you will notice that the error bar drastically increases. The reason for this is small statistics: there are only $2l + 1$ spherical harmonics for a given l. Physically this means that we cannot measure a lot of unrelated large scale fluctuations from our one observation point in the universe and thus, statistically speaking, our sample size is limited. The phenomenon is referred to as cosmic variance (cf. the remark in Sect. 10.3.1).

(c) Calculate the cross-over time t_{mv} from matter to vacuum energy dominance defined by $\rho_v(t_{mv})/\rho_m(t_{mv}) = 1$. Provide t_{mv} for both models in part (a).

2. In the context of (10.54) it is stated 'Note that the ratio ρ_b/ρ_γ does depend on time. It is much smaller than one right after inflation, but at the surface of last scattering it is already close to one'. What is the actual value of ρ_b/ρ_γ at the surface of last scattering (use the value of the redshift given in (10.43)).

3. (a) Assuming matter dominance, i.e. $\rho = \rho_m$, express the comoving horizon $l(t_o = t_{today}, t_e = 0)$ and the attendant physical or proper horizon $D_{h,p}(t_o = t_{today}, t_e = 0)$ in terms of $a(t_o)$ and/or $H(t_o)$.

   (b) Use our standard model, i.e. $\Omega_v = 0.7$, $\Omega_K = 0$, $\Omega_m = 0.3$ and $\Omega_r = 9 \cdot 10^{-5}$, to calculate $D_{h,p}(t_o = t_{today}, t_e = 0)$ in lt yr. Compare your number to the suspected patch size D_o in Sect. 11.2. Also compare your number with the estimate $D_{h,p} \approx 1/H(t_o = t_{today})$.

4. Suppose that astronomers measure the age of a galaxy with redshift $z = 2.5$. How old would this galaxy have to be, at the time the light from it was emitted, in order to rule out the hypothesis that $\Omega_m = 1$ with all other Ωs being zero? Use $H_0 = 70\,\text{km}/(\text{s Mpc})$.

5. Suppose that $\Omega_m = 0.26$ and $\Omega_v = 0.74$, with the other Ωs being zero. What is the redshift at which the expansion of the universe stopped decelerating and began to accelerate?

6. Derive (10.42).

7. Discuss the relation $T(z) = T_o(1 + z)$, where T_o is the present temperature.

8. In Problem 2 of Chap. 6 we have investigated the Schwarzschild-de Sitter metric which describes a single point mass in a universe with a cosmological constant.

   (a) Let us take the physical value of the cosmological constant and assume that the point mass is our Sun. Is the approximation $MG\sqrt{\Lambda} \ll 1$ justified?

   (b) What is the largest stable circular orbit around the Sun? How does it compare to the size of the solar system?

   (c) Now let us apply the same model to our local group of galaxies with a total mass of $\sim 2 \times 10^{12}$ solar masses. Is the approximation $MG\sqrt{\Lambda} \ll 1$ still justified? What is the radius of the largest circular orbit? How does it compare to the size of the local group of ~ 2–3 Mpc?

Chapter 11
Inflation

The classical Big Bang cosmology that we have explored in the last few chapters poses some fundamental questions, most notably the uniformity of the cosmic background radiation and the overall flatness of space. In this chapter we develop the basics of inflationary models, which answer these questions at the expense of introducing a hypothetical scalar inflaton field.

11.1 Why Inflation?

Let's remind ourselves what a cosmological constant does in Einstein's equations. It is an energy density, which does not dilute. In the special case of vacuum energy and flat space

$$\left(\frac{\dot{a}}{a}\right)^2 = \frac{8\pi}{3} G \rho_v = H^2 . \tag{11.1}$$

Hence

$$a = e^{Ht} \tag{11.2}$$

and

$$d\tau^2 = dt^2 - e^{2Ht}(dx^2 + dy^2 + dz^2) . \tag{11.3}$$

This is called de Sitter space. It is an important spacetime in cosmology for two reasons: First it describes the future of our universe, which will be ever more dominated by the cosmological constant. Second it might have appeared in a very early epoch

© Springer Nature Switzerland AG 2020

R. Hentschke and C. Hölbling, *A Short Course in General Relativity and Cosmology*, Undergraduate Lecture Notes in Physics,
https://doi.org/10.1007/978-3-030-46384-7_11

of our universe, before it became radiation dominated. The exponential expansion
back then is conjectured to have happened much faster and then stopped. This general idea is known as inflation. But why did this idea get any traction and where did
it come from? Essentially there are two peculiarities about the initial conditions of
our universe that inflation can explain, namely the flatness problem and the horizon
problem.

The flatness problem is the question why our universe today is spatially flat, i.e.
why the curvature term Ω_K as defined in (9.24) is compatible with 0. To see why this
is a problem, we can look at the first Friedmann equation in the form (9.26). It tells
us that the contribution of matter and radiation to the energy density of the universe
drops as $\propto a^{-3}$ and a^{-4} with the scale factor a while the curvature contribution
only goes down as $\propto a^{-2}$. Thus, relative to matter the curvature contribution grows
$\propto a$ while relative to radiation it even increases $\propto a^2$. But despite this huge relative
growth the curvature contribution is not even detectable today, so it must have been
extremely small in the early universe. This is the flatness problem and if we look again
at (9.26) we see how inflation can solve it: The vacuum energy density stays constant,
so relative to it the curvature term is suppressed by a factor a^2. Inflation increases
the scale factor exponentially and if it keeps going through enough e-foldings, the
curvature term is exponentially suppressed. Current estimates suggest a minimum
number of about 65 e-foldings of the universe's size during inflation to explain the
degree of flatness within current observational limits.

The horizon problem, which we have already discussed in Sect. 9.3, states that
the far ends of the visible universe must have been in causal contact before the time
of last scattering, because the surface of last scattering is already pretty smooth (cf.
Fig. 10.5). To see that this problem is also solved by an early inflationary epoch we
look at the causal horizon in the de Sitter metric (11.1). Imagine that an observer
sends out a light ray in the x-direction. How far will it travel in a small time interval
dt in comoving coordinates? Since it is a light ray, $d\tau = 0$ and thus

$$dx = e^{-Ht} dt . \tag{11.4}$$

Integrating the relation and assuming we sent out the light ray at $t = t_e$ and $x = 0$,
we obtain
$$x = \frac{e^{-Ht_e} - e^{-Ht}}{H} \tag{11.5}$$

so the light ray actually never progresses further than a distance $x_{max} = \exp$
$(-Ht_e)/H$, even if it travels forever! In addition, the maximal physical distance
it can reach is $1/H$ (cf. (10.26)), which means that regions that were causally connected become disconnected as time progresses.

The original motivation of the inventor of inflation, Alan Guth,[1] had to do with the
apparent scarceness of magnetic monopoles, which are generic predictions of grand

---

[1] A nice personal account of this can be found in his book entitled *The Inflationary Universe*. We
also recommend his e-course (Inflationary Cosmology: Is Our Universe Part of a Multiverse; http://
ocw.mit.edu/8-286F13).

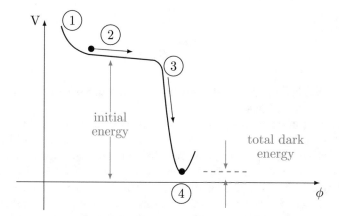

Fig. 11.1 A scenario of inflation in terms of the inflaton potential $V(\phi)$: The scalar field ϕ 'rolls' down the potential from an unknown origin (1). During its trip along the almost flat part (2) the universe inflates rapidly. This phase ends at an 'edge' (3), where the potential drops to the current value of the vacuum or dark energy, thereby producing the conditions which are the starting point of the classical hot Big Bang (4)

unified theories. In grand unified theories these monopoles have too large a mass to be produced in the lab, but they should have been produced in the early universe when temperatures were high enough, just like any other species of particles. Their non-observation thus would rule out grand unified theories if there is no mechanism to dilute them. Inflation offers such a mechanism, provided that it happens late enough, i.e. after the monopoles have frozen out.

11.2 Inflation and the Potential of Vacuum

If we want inflation to happen, we need a vacuum energy that must have been vastly greater than it is today. Then, at some point, it must have dropped to almost zero. While eliminating the curvature density in the process almost entirely it must have left a substantial radiation and matter density. How did this happen?

The basic idea is to introduce a new, hypothetical scalar field, the so called infla-ton field[2] ϕ. The inflaton field is described by a quantum field theory, but we can understand the most important points by a mostly classical description. The inflaton field has a potential $V(\phi)$ depicted roughly in Fig. 11.1. Note that

$$\rho_v \approx V(\phi) \tag{11.6}$$

---

[2]Among the many specific models of inflation, there are some that identify the inflaton field with the only known fundamental scalar field, the Higgs. Although technically the inflaton is not a new field in these models, it still requires nonstandard couplings and thus hypothetical 'new physics'.

and therefore

$$H \approx \sqrt{\frac{8\pi}{3} G V(\phi)} \ . \tag{11.7}$$

Also note that ϕ in principle can depend on position, which means that the expansion is different in different places.

Here is what observations point to: Point 1 in Fig. 11.1 is an unknown starting point. In particular we don't know how we got there. The following inflation period consists of roughly 65 e-foldings, during which the universe expanded by a factor e^{65}. The 65 e-foldings occur along the very shallow decline of the inflaton potential indicated by 2 in Fig. 11.1. It is not really clear how high the plateau is or how long it is. At point 3 in the same figure the field fell down the potential cliff from the false vacuum state[3] to where the universe is now (number 4 in Fig. 11.1). The energy gained by this rapid drop from the false vacuum caused a rise in temperature of the universe and filled it with radiation and matter, which usually is referred to as reheating. This really is the classical Big Bang. Before the drop, the universe was cold by comparison and the energy was essentially stored in the false vacuum.

This also gives us an idea how high the plateau should be.[4] If we want to obtain energies large enough for grand unification, the energy scale of which is $E_{\text{GUT}} \approx 10^{16}$ GeV, then

$$\rho_v \approx \frac{E_{\text{GUT}}^4}{\hbar^3 c^5} \approx 2.3 \cdot 10^{84} \text{ kg/m}^3 \ . \tag{11.8}$$

This then means that the ratio of ρ_v at the time of inflation and what it is now, i.e. $\rho_{v,0} \approx 0.7 \rho_{c,0}$, is about

$$\frac{\rho_v}{\rho_{v,0}} \approx 10^{110} \ ! \tag{11.9}$$

Using the energy density (11.8) we find that during inflation, i.e. on the plateau,

$$H^{-1} \approx 2.8 \cdot 10^{-38} \text{ s} \ . \tag{11.10}$$

This means that the 65 e-foldings took the time Δt given by

---

[3]The model of inflation originally suggested by A. Guth was based on a $V(\phi)$, exhibiting a local valley, i.e. a false vacuum state, at $\phi = 0$ from which the field escapes via quantum tunneling (the underlying theory was worked out mainly by S. Coleman and collaborators (see for instance Chap. 6 in [34])). However, the tunneling happens locally rather than globally and inflation ends in an 'heterogeneous mess', which is called the graceful exit problem. The old inflation model was superseded by the slow roll scenario of new inflation (mainly developed by A. Linde, P. Steinhardt, and A. Albrecht). This is essentially what we discuss here. In this scenario the false vacuum, in the sense of temporary, is not a local minimum of $V(\phi)$. In particular new inflation solves the graceful exit problem.

[4]In this paragraph we explicitly include the speed of light c into the formulas.

$$\frac{a(\text{end of inflation})}{a(\text{beginning of inflation})} = e^{65} \approx e^{H \Delta t} \ , \tag{11.11}$$

i.e.

$$\Delta t \sim 10^{-36} \text{ s} \ . \tag{11.12}$$

The horizon distance on the plateau was

$$D_h = c/H \approx 8 \cdot 10^{-30} \text{ m} \ . \tag{11.13}$$

By the end of inflation, a patch of universe of this size would have grown to

$$D = e^{65} \, 8 \cdot 10^{-30} \text{ m} \sim 10 \text{ cm} \ . \tag{11.14}$$

How large is D at the present time? Using $D \propto T^{-1}$ we have

$$D_o \approx \frac{10^{16} \text{ GeV}}{k_B \, 2.7 \text{ K}} \times 10 \text{ cm} = 450 \cdot 10^9 \text{ lt yr} \ . \tag{11.15}$$

This is bigger than what we need to explain our observations of the present universe. Note that all this developed from a tiny patch the size of the above D_h, where things were in contact. This means that the horizon problem has disappeared!

How about the other problem, the flatness problem? The key equation is the Friedman equation in the form

$$H^2 = \frac{8 \pi G}{3} \rho_v - \frac{K}{a^2} \ . \tag{11.16}$$

While a grows by a factor of $e^{65} \approx 10^{28}$, the first term on the right hand side remains basically constant. Thus by the end of inflation the second term on the right hand side will have diminished by a factor of 10^{-56}. This means that whatever it was before inflation, after inflation it has to very good approximation disappeared.[5]

What about the monopole problem? We believe this is more subtle, since it only appears on a theoretical level in grand unified theories, which are themselves hypothetical. The fact that no magnetic monopole has been observed yet is one of many constraints that need to be considered when constructing a specific theory of grand unification and inflation.

One might think that if ϕ 'rolls' down the cliff in Fig. 11.1 it would swing back and forth for all times. This is not so. But what then is the form of the necessary

---

[5]Note that the matter and radiation densities during inflation have diminished even faster, as $\propto a^{-3}$ and $\propto a^{-4}$. This is however not a problem, because all the matter and radiation we see in our universe today originated during the huge rise in temperature when the inflaton field fell down the cliff.

dissipation? It has something to do with the expansion of space. Let's look at the total energy density of the scalar field, i.e.

$$\rho_v = \frac{1}{2}\partial_\mu\phi\,\partial^\mu\phi + V(\phi)$$

$$= \frac{\dot\phi^2}{2} - \frac{\vec\nabla\phi^2}{2} + V(\phi)\,. \tag{11.17}$$

Note that we neglect the $(\partial_\mu\phi)^2$-terms in (11.6) and (11.7). Neglecting spatial variations of ϕ for the moment we can write for the Lagrangian of the inflaton field

$$\mathcal{L} = a^3(t)\left[\frac{\dot\phi^2}{2} - V(\phi)\right]\,. \tag{11.18}$$

Note that the factor $a^3(t)$ arises because the quantity in square brackets is a density. Equation (11.18) yields the Euler-Lagrange equation

$$\frac{d}{dt}a^3(t)\dot\phi = -a^3(t)\frac{\partial V}{\partial\phi}\,. \tag{11.19}$$

After calculating the time derivative we have

$$\ddot\phi + 3H\dot\phi = -\frac{\partial V}{\partial\phi}\,. \tag{11.20}$$

The second term on the left hand side is called Hubble friction. Of course, here we assume that the time dependence of H is weaker than the time dependence of ϕ.

At this point it looks as if we have overdone it. The exponential expansion has completely fattened the universe. If this is the case, how do we get any structure to form?

11.3 Cosmic Background Anisotropy and Structure Formation

Figure 11.2 illustrates the approach of the scalar field ϕ as it 'rolls' slowly towards the edge of the cliff as shown in Fig. 11.1 (cf. point 2). The vertical axes represent ordinary space. If ϕ was completely homogeneous it would 'fall over the edge' everywhere in space at the same time. However, ϕ is not completely homogeneous because of quantum fluctuations, here indicated by the wiggly red lines. As a consequence, ϕ in certain places in Fig. 11.2 (shaded in blue and labeled 1) goes over the edge before it does in other places (green shading labeled 2). Let's distinguish this by $\phi(1)$ and $\phi(2)$. Space in the 1-places continues to inflate more slowly in comparison to the

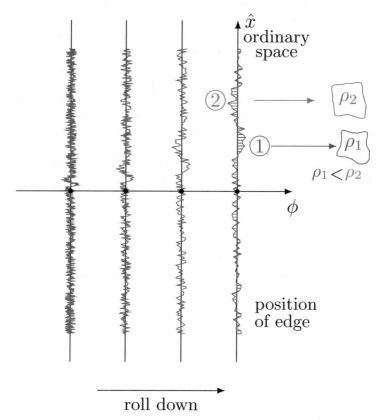

quantum fluctuations

Fig. 11.2 From quantum fluctuations to structure formation: The scalar field approaches the edge of the cliff along the horizontal direction. Wiggly red lines illustrate quantum fluctuations expanding in space as the cliff is approached. In certain places the scalar field goes over the edge sooner (1) or later (2). The result are underdense and overdense regions at the bottom of the cliff, developing according to their gravitation

2-places, where inflation remains fast. This gives rise to energy inhomogeneities. Note that the 1-places, where $\phi(1)$ has gone over the edge and the vacuum energy was 'dumped' into ordinary energy (photons, etc.), the energy density ρ_1 will dilute due to expansion. In the 2-places, where $\phi(2)$ has not yet reached the edge, this dilution does not yet happen, because the vacuum energy does not dilute while space expands. When $\phi(2)$ reaches the edge and converts its vacuum energy into ordinary energy, the energy density in the 2-places, ρ_2, will be greater than ρ_1. This difference is more pronounced the slower ϕ approaches the edge.

Note that there is a subsequent development where the overdense region gravitationally attracts additional matter from the underdense region so that the difference becomes more pronounced. This essentially is the origin of galaxies and their distribution.[6] Note however that the initial density variations are rather small as we can infer from the small fluctuations in the cosmic background radiations. This in turn imposes certain constraints on the flatness and the height of the plateau in Fig. 11.1.

We must be aware that the anisotropies in the cosmic microwave background are contributed by a number of different sources. The gravitational redshift or blueshift due to fluctuations in the gravitational potential at last scattering, which is what we have discussed here, is only one of them. In addition, there are intrinsic temperature fluctuations in the plasma or the Doppler effect due to velocity fluctuations in the plasma at last scattering and more. A in depth discussion is given in [26] beginning with Sect. 2.6.

When did the structure formation really take off? It happened around the time when radiation became unimportant. This has to do with pressure, i.e. with the equation of state (9.14). For radiation $\omega = 1/3$ and positive pressure homogenises things. Matter dominated means $\omega = 0$ and thus $P = 0$. Now gravity overcomes pressure and things start to cluster.

Nevertheless, the primordial fluctuations are still imprinted on the microwave background as shown in Fig. 10.5. It is perhaps the greatest success and evidence for inflation that the fluctuation pattern in the microwave background can be calculated quantitatively starting from the quantum fluctuations of the inflaton field on its way down to the edge in Fig. 11.1.

The picture we have developed in Fig. 11.2 is not complete yet. We return to the Lagrangian (11.18) and include the spatial variation (along the x-direction), i.e.

$$\mathcal{L} = a^3(t) \left[\frac{\dot{\phi}^2}{2} - \frac{\partial_x \phi^2}{2} - V(\phi) \right] . \tag{11.21}$$

Now instead of (11.20) we have

$$\ddot{\phi} + 3H\dot{\phi} - \frac{1}{a^2(t)} \partial_x^2 \phi \approx 0 . \tag{11.22}$$

Here we neglect the V-term, which is small, and does not affect the argument we want to make. We insert the solution ansatz

$$\phi \sim \phi_k(t) e^{ikx} \tag{11.23}$$

for one particular mode k, which yields

$$\ddot{\phi}_k(t) + 3H\dot{\phi}_k(t) + \frac{k^2}{a^2(t)} \phi_k(t) \approx 0 . \tag{11.24}$$

---

[6]Quantum fluctuations as the possible origin of structure in the universe is an idea that goes back at least as far as to a paper by Sakharov in 1965 [35].

This is just a simple damped harmonic oscillator. Remembering the discussion of the damped harmonic oscillator from classical mechanics (e.g. [4], p. 157) we conclude that there is no periodic motion when

$$\frac{3}{2}H > \frac{k}{a(t)} . \tag{11.25}$$

What is the physical implication of this inequality? Remember that according to (10.19) the distance to the horizon in a vacuum dominated universe is $D_h = 1/H$. Inserting this into (11.25) yields

$$\tilde{\lambda} > D_h , \tag{11.26}$$

where $\tilde{\lambda} = a(t)\lambda$. Note that λ is the wavelength of this particular fluctuation mode in coordinate space, which is not altered by the expansion of space. What is altered is $\tilde{\lambda}$. Nevertheless, for every λ, as space expands rapidly, comes the time when the inequality (11.26) is satisfied. The periodic motion of this mode ends—it is 'frozen' at a particular amplitude. We can interpret (11.26) by saying that the wave consists of pieces, roughly the length of D_h, which obviously can no longer communicate and thus 'do no longer know' that they belong to an oscillating wave. The wave therefore stops oscillating.

This process is going on for all modes k of the quantum fluctuations of the inflaton field as it rolls towards the edge in Fig. 11.1. The originally wild quantum fluctuations become frozen long-range variations, which pile on top of each other. Over time a structural 'steady state' develops, which is called a scale invariant spectrum. In the end this leads to comparisons with experimental measurements as shown in Fig. 11.3. We may also say that the emerging structure is characterised by a certain fractal dimension. Figure 11.4 is a cartoon, based on a certain fractal, which might serve to illustrate this approach to self-similarity on all scales.

Note that the wiggly red lines in Fig. 11.2 are to be understood as illustrations of this scale invariant frozen fluctuation structure. Note also that because $V(\phi)$ is large on the plateau and very small after the drop over the edge, H will also decrease dramatically. This in turn will vastly increase D_h, which means that the condition (11.26) for freezing no longer holds and the fluctuations come back to life. Nevertheless, as already stated, the remnants of the (almost) scale invariant spectrum are imprinted on the microwave background.

Can we see more clearly how the quantum fluctuation can lead to detectable features? Let's consider the following simplified line of reasoning. We start from a metric like (7.9), i.e.

$$d\tau^2 = dt^2 - a^2(t)e^{2\zeta(t,x)}dx^2 , \tag{11.27}$$

where $\zeta(t, x)$ is a curvature perturbation. On the plateau we have

$$a \sim e^N , \tag{11.28}$$

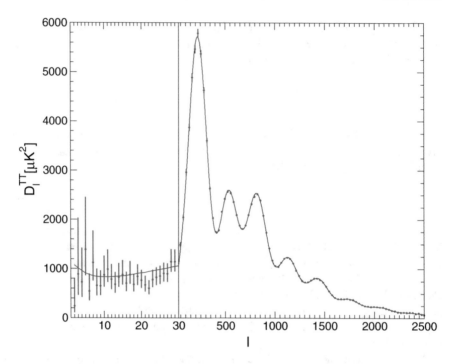

Fig. 11.3 The Planck temperature power spectrum. The black line corresponds to the best fit of the standard cosmological ΛCDM model. Note that modes $l > 30$ are binned for better visibility and the scale of the x-axis differs for them. Data and fits courtesy of the Planck Public Data Release 2 (https://irsa.ipac.caltech.edu/data/Planck/release_2/ancillary-data/). The results are presented in [18, 30]

where N is the number of e-foldings. Including a perturbation of the aforementioned type means that N is replaced by $N = \bar{N} + \delta N$. Thus

$$\zeta \sim \delta N = \delta \ln a = H \delta t$$
$$= \frac{H}{\dot{\phi}_k} \delta \phi_k \overset{*}{\sim} \frac{H^2}{\dot{\phi}_k} \overset{**}{\sim} \sqrt{\frac{\rho_k}{\epsilon}} . \qquad (11.29)$$

Here * and ** require explanations: (*) The quantity $\delta \phi_k$ is a measure for the width of the fluctuation in k-space. The assumption $\delta \phi_k / H \sim 1$ is akin to the uncertainty relation in quantum mechanics, because $1/H$ is the attendant localisation in position space.[7] (**) We introduce the so called equation of state parameter defined via[8]

$$\epsilon = \frac{3}{2}(1 + \omega). \qquad (11.30)$$

---

[7] More precisely: $\sqrt{\langle \delta \phi^2(k) \rangle} \sim H$, where $\langle ... \rangle$ is a quantum average.

[8] Note that ϵ is a constant in the following derivation!.

Fig. 11.4 Cartoon of the development of a scale invariant spectrum. A mode freezes when condition (11.25) is met as indicated by the red circle. Space expands and the process repeats

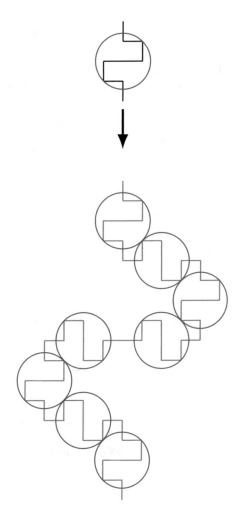

Note that $\omega = P/\rho$ (cf. (9.14)).[9] Here

$$P = \frac{1}{2}\dot{\phi}^2 - V \quad \text{and} \quad \rho = \frac{1}{2}\dot{\phi}^2 + V .$$
(11.31)

The formula for ρ is perhaps more obvious than the one for P, which follows via $\rho = \frac{1}{2}\dot{\phi}^2 + V$ combined with (8.34) and (11.20). A derivation involving the energy-momentum tensor of the scalar field can be found in [26] (Appendix B12). Thus

---

[9] We had argued that the vacuum energy density is constant and $\omega = -1$. This means $\epsilon = 0$. However, the vacuum energy density is not exactly constant due to the slight slope built into the potential V.

$$\frac{2}{3}\epsilon = \frac{\dot{\phi}_k^2}{\rho_k} \; , \tag{11.32}$$

which completes (11.29).

Next we combine (8.34), the equation of state (9.14) and (11.30) which yields

$$\dot{\rho}_k = -2\epsilon\rho_k\frac{\dot{a}}{a} \; . \tag{11.33}$$

Hence

$$\frac{\delta\rho_k}{\rho_k} \sim -2\epsilon H\delta t \stackrel{(11.29)}{\sim} -\sqrt{\epsilon\rho_k} \; . \tag{11.34}$$

Our above reasoning in the context of the temperature variations in the CMB lets us expect an initial power law form like

$$\frac{\delta\rho_k}{\rho_k} \sim k^{(n_s-1)/2} \; . \tag{11.35}$$

Comparing the last two equations yields

$$\ln\epsilon\rho_k \approx (n_s - 1)\ln k \tag{11.36}$$

or

$$\mathrm{d}\ln\rho_k \approx (n_s - 1)\mathrm{d}\ln k \; . \tag{11.37}$$

Note that on the plateau according to (11.25) we have[10] $a \sim k$ and thus $\ln k \sim \ln a = N$, i.e.

$$\mathrm{d}\ln\rho_k \approx (n_s - 1)\mathrm{d}N \; . \tag{11.38}$$

Using once again (11.34) in conjunction with $H\mathrm{d}t = \mathrm{d}N$, we obtain[11]

$$n_s \approx 1 - 2\epsilon. \tag{11.39}$$

We find the relation between ϵ and $N = N_{\max}$, the total number of e-foldings, by integrating (11.33), which yields $\rho_k \sim e^{-2N\epsilon}$. During the slow roll and the many e-foldings ρ does not change much, no matter how big the final N is, and thus we conclude, most naturally,

---

[10]The k we are talking about here is the one when (11.25) is given by $3H/2 \approx k/a$—the 'just freezing' mode. And because H is nearly constant on the plateau we infer $a \approx k$ for this mode.

[11]The quantity $1 - n_s$ is called the tilt of the spectrum.

$$\epsilon = 1/N . \tag{11.40}$$

This then yields the spectral index[12]

$$n_s \approx 1 - 2/N \approx 0.969 \text{ with } N = 65 \tag{11.41}$$

Note that $n_s = 1$ in the limit $N \to \infty$. This is the limit of scale invariance or self-similarity of the fluctuation spectrum.

In order to really understand the meaning of n_s, we must relate (11.35) to the quantity Δ_T depicted in Fig. 10.7. As the first step we interject a box in which we show how to calculate the coefficients C_l from the Fourier transform of an attendant pair correlation function of the density fluctuations.

• **Calculating $l(l + 1)C_l$ from the Fourier Transform of Attendant Pair Correlation Functions:** We begin with a short reminder of Fourier transformation formulas, i.e.

$$A(\vec{r}) = \frac{1}{(2\pi)^3} \int d^3k A_{\vec{k}} e^{-i\vec{k}\cdot\vec{r}} \quad \text{and} \quad A_{\vec{k}} = \int d^3r A(\vec{r}) e^{i\vec{k}\cdot\vec{r}} \tag{11.42}$$

as well as

$$\delta(\vec{k} - \vec{k}') = \frac{1}{(2\pi)^3} \int d^3r e^{-i(\vec{k}-\vec{k}')\cdot\vec{r}} . \tag{11.43}$$

Let's suppose the quantity $A(\vec{x}_o, \vec{e}, \eta)$ is measured by an observer at $\vec{x}_o$. The line of sight of the observer is defined by the unit vector $\vec{e}$ and η is the distance from the observer to where the observed signal is emitted. Expressing $A(\vec{x}_o, \vec{e}, \eta)$ via its Fourier transform yields

$$A(\vec{x}_o, \vec{e}, \eta) = \int \frac{d^3k}{(2\pi)^3} A_{\vec{k}} e^{-i\vec{k}\cdot(\vec{x}_o + \vec{e}\eta)} . \tag{11.44}$$

We want to calculate the pair or two-point correlation function

$$C(\gamma) = \langle A(\vec{e}_1, \eta_1) A(\vec{e}_2, \eta_2) \rangle , \tag{11.45}$$

where $\cos\gamma = \vec{e}_1 \cdot \vec{e}_2$. The similarity to (10.27) is quite obvious. In addition we can express A in terms of its Fourier components $A_{\vec{k}}$ in $\vec{k}$-space. This is the space we have used to develop the formalism which led to n_s. Thus, we are

---

[12]The spectral index n_s was calculated first by V. F. Mukhanov and G. V. Chibisov in 1981 [36] (cf. [37]). The interesting history of quantum fluctuations in cosmology is described in [38].

en route from the $\vec{k}$-space representation of a quantity to the description of its correlations on the surface of a sphere.

Note that the observer position $\vec{x}_o$ in (11.45) is missing. This is because we assume a homogeneous space in which the observer's position has no special meaning and thus we can use it for averaging, i.e.

$$C(\gamma) = \frac{1}{V} \int_V d^3 x_o A(\vec{x}_o, \vec{e}_1, \eta_1) A(\vec{x}_o \vec{e}_2, \eta_2) \ . \tag{11.46}$$

Replacing the $A(\vec{x}_o, \vec{e}_i, \eta_i)$ in the above equation by their Fourier transforms and carrying out most integrations yields

$$C(\gamma) = \frac{1}{2\pi^2 V} \int_V dk k^2 \langle A_{\vec{k}} A_{-\vec{k}} \rangle \frac{\sin(k|\vec{e}_1 \eta_1 - \vec{e}_2 \eta_2|)}{k|\vec{e}_1 \eta_1 - \vec{e}_2 \eta_2|} \tag{11.47}$$

(Problem 2). For the final result we need the identity

$$\frac{\sin \lambda R}{\lambda R} = \sum_{l=0}^{\infty} (2l+1) j_l(\lambda r) j_l(\lambda \rho) P_l(\cos \gamma) \ , \tag{11.48}$$

where j_l is a spherical Bessel function of the first kind and $R = \sqrt{r^2 + \rho^2 - 2r\rho \cos \gamma}$ (e.g. [39] (10.1.45)). Combination of the identity with (11.47) yields

$$C(\gamma) = \sum_{l=0}^{\infty} \frac{2l+1}{4\pi} C_l P_l(\cos \gamma) \ ,$$

which is identical to (10.30), and

$$\boxed{C_l = \frac{2}{\pi} \int_V dk k^2 \frac{\langle A_{\vec{k}} A_{-\vec{k}} \rangle}{V} j_l^2(k\eta)} \ . \tag{11.49}$$

In this equation we assume $\eta \equiv \eta_1 = \eta_2$, which we justify below (**).

Before we leave this box let us just assume a power-law form for $\langle A_{\vec{k}} A_{-\vec{k}} \rangle$, e.g.

$$\frac{\langle A_{\vec{k}} A_{-\vec{k}} \rangle}{V} = Bk^{-4+n} \ , \tag{11.50}$$

and compute the integral in (11.49). Thus, with $s = \eta k$, we find

$$C_l = B\eta^{1-n} \frac{2}{\pi} \int_0^\infty ds\, s^{n-2} j_l^2(s)$$

$$= B\eta^{1-n} \frac{1}{\sqrt{4\pi}} \frac{\Gamma\left(l + \frac{n-1}{2}\right) \Gamma\left(\frac{3-n}{2}\right)}{\Gamma\left(l + \frac{5-n}{2}\right) \Gamma\left(\frac{4-n}{2}\right)} \tag{11.51}$$

(the integral may be found for instance in [40] (6.574) or (today) it may be obtained via a suitable program like *Mathematica*). In general C_l will depend on l, but for the special case $n = 1$ we find

$$l(l+1)C_l = \frac{B}{\pi} \quad (n = 1) . \tag{11.52}$$

The significance of the formula is that $n = 1$ yields a constant Δ_T in (10.32) and thus results in no structural features at all for any angular range. The underlying power spectrum $\langle A_{\vec{k}} A_{-\vec{k}} \rangle$ is scale invariant.

(**) The quantities η_i here are the comoving distances between the observer and the emitter (cf. (10.21) or (10.2)). Both photon emitters $i = 1$ and $i = 2$ are on the surface of last scattering and thus $\eta \equiv \eta_1 = \eta_2$ is justified.

By the mechanism discussed above inflation will generate a primordial or initial power spectrum, i.e. the Fourier transform of the primordial density-density correlation functions $P_i(k) = \langle \rho_{\vec{k}} \rho_{-\vec{k}} \rangle$.[13] The latter is reprocessed by a number of physical effects that occur at latter times. In the previous chapter we have mentioned some of them when we discussed features in the CMB. The reprocessing is quantified in a transfer function $T(k)$, which yields the power spectrum at a latter time, i.e.

$$P(k) \sim T(k) P_i(k) . \tag{11.53}$$

When we observe the structures in the CMB that time will be the time of recombination. In order to get from the density fluctuations to temperature fluctuations, we 'convert' density ρ into gravitational potential ϕ with the help of the Poisson equation. Writing the latter in $\vec{k}$-space means $k^2 \phi_k \sim \rho_k$. Therefore the power spectrum $P_\phi(k)$ of the potential is related to $P(k)$ via

$$P_\phi(k) \sim k^{-4} P(k) . \tag{11.54}$$

This also is the reason for the ominous power -4 in the example (11.50). Potential fluctuations in turn can be expressed in temperature fluctuations as we had discussed in the context of (10.34), i.e. $P_\phi(k) \sim \langle \delta T_{\vec{k}} \delta T_{-\vec{k}} \rangle$, completing the roadmap from density fluctuations to the observation of temperature fluctuations correlations in the CMB.

Returning to the spectral index in (11.41) we conclude that our best chance for a concrete comparison of this prediction of inflation with the CMB spectrum is in the

---

[13]Remark: In a different field the power spectrum might be called structure factor.

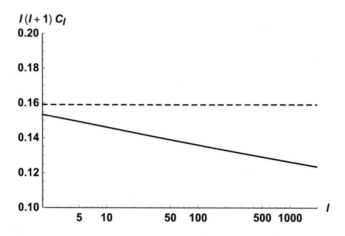

Fig. 11.5 $l(l + 1)C_l$ versus l calculated with (11.51) using $B = \eta = 1$, for $n = 1$ (dashed line) and $n = n_s = 0.968$ (solid line)

plateau-like part of Fig. 11.3, where $l \lesssim 30$ and the primordial density structure is essentially preserved. Note that we do get a plateau from (11.49) if we use the power law (11.50) with $n = 1$ or

$$P(k) \sim k . \tag{11.55}$$

This is known as the scale invariant or Harrison-Zeldovich spectrum. The prediction of inflation is that the finite number of e-foldings of the universe produced a slightly smaller power given by the spectral index in (11.41). Had inflation continued, $N \rightarrow \infty$, the result would have been $n_s = 1$ and the structure of the then frozen fluctuations would have been truly scale invariant. The quantity $\delta\rho_k$ therefore refers to the deviation from scale-invariance due to a finite number of e-foldings.

Figure 11.5 compares $l(l + 1)C_l$, calculated with (11.51) using $B = \eta = 1$, for $n = 1$ and $n = n_s = 0.968$.[14] This shows that inflation predicts a slightly tilted plateau in comparison to the scale invariant spectrum. This tilt is also seen in the fit to the Planck data at low l in Fig. 11.3. Note that the model of inflation, which we have discussed here, is not the only one. Other models produce also slightly other values for n_s. But, as is seen in Fig. 11.3, there is considerable scatter in the data at low l, which makes a distinction been these models difficult. We like to add that the preceding discussion is presented with much more accuracy and detail in Mukhanov's book ([41]; Chap. 9).

However, n_s is only one of a number of quantities, which have been calculated from inflation models and compared to observations of the microwave background. Thus, inflation has been successful in a number of ways, i.e. it solves problems of the classical Big Bang cosmology and it makes predictions which can be checked experimentally. But there are problems—perhaps a better expression is challenges—

---

[14]The latest value from the Planck collaboration is $n_s = 0.968 \pm 0.006$.

as well. There is what is called 'eternal inflation' accompanied by the 'multiverse' concept. The basic problem is that there are always some regions, however small they may be originally, where ϕ has not 'gone over the edge'. But then the regions which have not gone over the edge grow exponentially via e-foldings and overcome the over-the-edge-regions, which form 'bubble universes'. This hampers the predictive power of inflation, because in an infinite number of universes everything can happen!

Before leaving the subject let's ask the question when the downhill motion of the field is overcome by the quantum dispersion, which always also acts in the opposite direction, i.e. uphill. The classical downhill motion is given by

$$\delta_{cl}\phi = H^{-1}\dot{\phi} \overset{(11.20)}{\approx} -H^{-1}\frac{\partial V/\partial \phi}{3H} . \tag{11.56}$$

The quantum dispersion is

$$\delta\phi \sim H \tag{11.57}$$

(cf. above). Thus, if $\delta\phi/\delta_{cl}\phi > 1$ then the field moves uphill and continues to inflate. This only needs a sufficiently flat potential, because with $H^2 \sim V$ we have

$$\frac{\delta\phi}{\delta_{cl}\phi} \sim -\frac{V^{3/2}}{\partial V/\partial \phi} . \tag{11.58}$$

Eternal inflation is a problem of the 'future'. But there also is a problem with the past. It is difficult to construct a beginning of inflation. Starting the field from rest does not appear plausible due to the always present quantum fluctuations.

Inflation certainly has been an extremely inspirational and in many ways successful concept over the past 30 years.[15] However, we have seen that in spite of this inflation is posing numerous, yet unanswered questions as well.

11.4 Problems

1. Describe how Fig. 9.3 changes if we use conformal time $\eta = \int dt/a(t)$ along the 'time' axis and the coordinate radius r along the 'space' axis. Assume that the period following inflation is exclusively matter dominated. Consider inflation based on the numbers given in Sect. 11.2. Illustrate your reasoning by a sketch in the η-r-plane supported by some numbers.
2. Carry out the steps leading from (11.45) to (11.46).

---

[15] A very good text, covering inflation theory as well as earlier developments leading up to it, is the book by V. Mukhanov [41].

3. $C(\gamma)$ of the Harrison-Zeldovich spectrum: In Sect. 10.3.1 we discuss various simple trial functions for $C(\gamma)$ and the Δ_T they produce (cf. (10.28) and (10.29) as well as Fig. 10.6). How does $C(\gamma)$ look like in the case of the Harrison-Zeldovich spectrum.

   (a) Use (11.52) to plot the attendant $C(\gamma)$ (omit the $l = 0$ term) for $l_{max} = 10$ and the largest l_{max} you can handle on your computer. Use a log-scale for the γ-axis.

   (b) Obtain the real-space P_ϕ, i.e. $P_\phi(r)$, in the limit of small r for the Harrison-Zeldovich spectrum. Include your result, i.e. the leading r-dependence of $P_\phi(r)$ plus a suitably chosen constant and include this function in the plot of $C(\gamma)$ of part (a). Assume $\gamma \propto r$ for small r. Hint: Compute the Fourier transform of $P_\phi(k) = V B k^{-4+n}$, where $n = 1$. For the lower limit of the integration over k introduce the cutoff $k_{min} = 2\pi/L$, where L is the linear dimension of the volume V. Your result should be the sought after leading term, which is a function of $2\pi r/L$, plus a constant (in the limit $k_{min} \to 0$).

Appendix A
Constants and Units

Fundamental Physical Constants:

quantity	symbol	value	unit
speed of light	c	299792458	$\mathrm{ms^{-1}}$
Planck constant	h	$6.62606876 \cdot 10^{-34}$	Js
	$\hbar = \frac{h}{2\pi}$	$1.054571596 \cdot 10^{-34}$	Js
Gravitation constant	G	$6.673 \cdot 10^{-11}$	$\mathrm{m^3 kg^{-1} s^{-2}}$
Planck length $\left(\hbar G/c^3\right)^{1/2}$	l_P	$1.6160 \cdot 10^{-35}$	m
Planck time l_P/c	t_P	$5.3906 \cdot 10^{-44}$	s
Planck mass $l_P c^2/G$	m_P	$2.1765 \cdot 10^{-8}$	kg
elementary charge	e	$1.602176462 \cdot 10^{-19}$	C
electron mass	m_e	$9.10938188 \cdot 10^{-31}$	kg
proton mass	m_p	$1.67262158 \cdot 10^{-27}$	kg
atomic mass unit	u	$1.66053873 \cdot 10^{-27}$	kg
Boltzmann constant	k_B	$1.3806503 \cdot 10^{-23}$	$\mathrm{JK^{-1}}$
Rydberg constant $\left(\alpha^2 m_e c/2h\right)$	R_∞	10973731.568	$\mathrm{m^{-1}}$

Source http://physics.nist.gov/constants

Energy equivalents:

conversion to J		
1 kg	$(1\,\mathrm{kg})\, c^2 =$	$8.987551787 \times 10^{16}$ J
$1\,\mathrm{m^{-1}}$	$(1\,\mathrm{m^{-1}})\, hc =$	$1.98644544\,(16) \times 10^{-25}$ J
1 Hz	$(1\,\mathrm{Hz})\, h =$	$6.62606876\,(52) \times 10^{-34}$ J
1 K	$(1\,\mathrm{K})\, k_B =$	$1.3806503\,(24) \times 10^{-23}$ J
1 eV	$(1\,\mathrm{eV}) =$	$1.602176462\,(63) \times 10^{-19}$ J
1 E_h	$(1\,E_h) =$	$4.35974381\,(34) \times 10^{-18}$ J

© Springer Nature Switzerland AG 2020
R. Hentschke and C. Hölbling, *A Short Course in General Relativity and Cosmology*, Undergraduate Lecture Notes in Physics,
https://doi.org/10.1007/978-3-030-46384-7

Astronomical Constants and Units:

quantity	symbol	value	unit
astronomical unit	1 AU	$1.496 \cdot 10^{11}$	m
Parsec	1 pc	$3.086 \cdot 10^{16}$	m
		3.26	light-years
		206265	AU
light-year	1 lt yr	$9.454 \cdot 10^{15}$	m
Earth			
radius	$R_{\oplus}$	$6.378 \cdot 10^6$	m
mass	$M_{\oplus}$	$5.973 \cdot 10^{24}$	kg
	$GM_{\oplus}/(R_{\oplus}c^2)$	$6.95 \cdot 10^{-10}$	
	$GM_{\oplus}/c^2$	0.443	cm
Moon			
radius	$R_{\mathbb{C}}$	$1.738 \cdot 10^6$	m
mass	$M_{\mathbb{C}}$	$7.35 \cdot 10^{22}$	kg
	$GM_{\mathbb{C}}/(R_M c^2)$	$3.14 \cdot 10^{-11}$	
	$GM_{\mathbb{C}}/c^2$	0.00545	cm
Sun			
radius	$R_{\odot}$	$6.96 \cdot 10^8$	m
mass	$M_{\odot}$	$1.989 \cdot 10^{30}$	kg
	$GM_{\odot}/(R_{\odot}c^2)$	$2.12 \cdot 10^{-6}$	
	$GM_{\odot}/c^2$	1.475	km
Milky Way			
sun - galactic center			
distance		8.5	kpc
rotation velocity		220	km s^{-1}
Hubble time	$1/H_0$	$\approx 4.4 \cdot 10^{17}$	s
		$\approx 14 \cdot 10^9$	y
Andromeda galaxy			
distance to		$2.5 \cdot 10^6$	light-years
linear dimension		$2.2 \cdot 10^5$	light-years

Appendix B
Geodesic Deviation

Here we derive an equation, allowing to calculate the separation variation between two geodesics at equal spaced points, even if they are parallel originally. Let's assume the point A in Fig. B.1 indicates a local inertial frame. In this case we have at A according to (2.49)

$$\frac{d^2 x^\alpha(A)}{d\tau^2} = 0 .$$ (B.1)

At A' we are outside the inertial frame at A and thus we must use (2.58), i.e.

$$\frac{d^2 x^\alpha(A')}{d\tau^2} + \Gamma^\alpha_{00}(A') = 0 .$$ (B.2)

Note that the points we connect via $\xi^\alpha(\tau)$ are at rest in their respective frames of reference. We can express $\Gamma^\alpha_{00}(A')$ via a Taylor expansion at A, i.e.

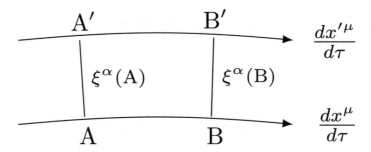

Fig. B.1 Two geodesics separated by distance $\xi^\alpha(\tau)$

© Springer Nature Switzerland AG 2020

R. Hentschke and C. Hölbling, *A Short Course in General Relativity and Cosmology*, Undergraduate Lecture Notes in Physics, https://doi.org/10.1007/978-3-030-46384-7

$$\Gamma^{\alpha}_{00}(A') = \underbrace{\Gamma^{\alpha}_{00}(A)}_{=0} + \Gamma_{00,\beta}(A)\xi^{\beta} .$$ (B.3)

Note that the first term vanishes, because A is a local inertial frame. Putting all three equation together and noting that $\xi^{\alpha} = x'^{\alpha} - x^{\alpha}$ we find

$$\frac{d^2\xi^{\alpha}}{d\tau^2} = -\Gamma^{\alpha}_{00,\beta}\xi^{\beta} .$$ (B.4)

Next we derive a second equation involving $\frac{d^2\xi^{\alpha}}{d\tau^2}$. Here we use the covariant derivative along the lower geodesic at A (cf. (3.60)):

$$
\begin{aligned}
\nabla_{\tau}\nabla_{\tau}\xi^{\alpha} &= \frac{d}{d\tau}\left(\nabla_{\tau}\xi^{\alpha}\right) + \Gamma^{\alpha}_{\beta 0}\left(\nabla_{\tau}\xi^{\beta}\right) \\
&= \frac{d}{d\tau}\left(\frac{d\xi^{\alpha}}{d\tau} + \Gamma^{\alpha}_{\beta 0}\xi^{\beta}\right) \\
&\quad + \underbrace{\Gamma^{\alpha}_{\beta 0}}_{=0}\left(\frac{d\xi^{\beta}}{d\tau} + \underbrace{\Gamma^{\beta}_{\nu 0}}_{=0}\xi^{\nu}\right) \\
&= \frac{d^2\xi^{\alpha}}{d\tau^2} + \Gamma^{\alpha}_{\beta 0,0}\xi^{\beta} .
\end{aligned}
$$ (B.5)

Note that $\Gamma^{\alpha}_{\beta 0}$ and $\Gamma^{\beta}_{\nu 0}$ vanish at A—again because it is a local inertial frame.

The combination of (B.4) and (B.5) yields

$$\nabla_{\tau}\nabla_{\tau}\xi^{\alpha} = \left(\Gamma^{\alpha}_{\beta 0,0} - \Gamma^{\alpha}_{00,\beta}\right)\xi^{\beta} \stackrel{(4.5)}{=} \mathcal{R}^{\alpha}_{00\beta}\xi^{\beta}$$ (B.6)

at A.

In principle this is the equation we need in our section on gravitational waves. But it is easy to wright down the frame independent form of the geodesic deviation:

$$\nabla_{\tau}\nabla_{\tau}\xi^{\alpha} = \mathcal{R}^{\alpha}_{\mu\nu\beta}\frac{dx^{\mu}}{d\tau}\frac{dx^{\nu}}{d\tau}\xi^{\beta} .$$ (B.7)

Appendix C
Temperature at Recombination

During the cosmic evolution a phase called recombination occurred. Neutral hydrogen and helium was formed, when the temperature had dropped to about 3000 K. Note that recombination may be misleading, because no neutral atoms had ever existed until this point. But this is the usual term and in addition recombination is still occurring today in the atmospheres of stars. In the following we still want to get a feeling for why it is associated with such a distinct temperature and why this temperature is close to 3000 K.

Here we study the reaction

$$p + e \rightleftharpoons 1s .$$

p and e stand for one proton and one electron, respectively, while $1s$ denotes the atomic hydrogen ground state. Using the mass action principle from chemistry applied to a gas phase reaction we may write

$$\frac{x_{1s}}{x_p x_e} = P K(T, P^o) \tag{C.1}$$

(cf. [22]; p. 114ff). The quantities x are mole fractions, P is the pressure and K is the equilibrium 'constant'. Note that the superscript o indicates a certain reference state. The quantity which we are interested in is the fraction of ionised hydrogen, i.e.

$$X = \frac{x_p}{x_p + x_{1s}} . \tag{C.2}$$

Combination of the last two equations yields

$$X(1 + SX) = 1 , \tag{C.3}$$

where

© Springer Nature Switzerland AG 2020
R. Hentschke and C. Hölbling, *A Short Course in General Relativity and Cosmology*, Undergraduate Lecture Notes in Physics,
https://doi.org/10.1007/978-3-030-46384-7

$$S = (\rho_p + \rho_{1s}) P K(T, P^o)/\rho \, . \tag{C.4}$$

Note that $x_p = x_e$ and $\rho_i = \rho x_i$, where ρ is the total number density of massive particles in the universe at this time. Equation (C.3) is the Saha equation (Meghnad Saha, 1893–1956, Indian astrophysicist).

In the following we want $X = X(T)$ and thus we need the explicit temperature dependence of $K(T, P^o)$. The latter quantity is given by

$$K(T, P^o) = \frac{1}{P^o} \exp\left[\frac{1}{RT}\left(\mu_p(T, P^o) + \mu_e(T, P^o) - \mu_{1s}(T, P^o)\right)\right] . \tag{C.5}$$

Note that $R = N_A k_B$, where N_A is Avogadro's number. The μ_i are the chemical potentials of the various components per mole. Their temperature dependence follows from the Gibbs–Helmholtz equation, i.e.

$$\left.\frac{\partial \mu_i(T, P, n_j)/T}{\partial T}\right|_{P,n_j} = -\frac{h_i}{T^2} \tag{C.6}$$

(cf. [22]; p. 114). Hence

$$\frac{\mu_i(T, P^o)}{T} = \frac{\mu_i(T^o, P^o)}{T^o} - \int_{T^o}^{T} dT' \frac{h_i(T')}{T'^2} \, . \tag{C.7}$$

The partial molar enthalpy is $h_i = e_i + RT$, where the internal energy is $e_i = e_i^{(o)} + 3RT/2$. We also use $e_{1s}^{(o)} - e_p^{(o)} - e_e^{(o)} = -13.6\,\mathrm{eV}\, N_A$, where the right side is the negative ionisation energy for one mole of 1s hydrogen. Overall we obtain

$$\frac{\mu_i(T, P^o)}{RT} = \frac{\mu_i(T^o, P^o)}{RT^o} + \frac{e_i^{(o)}}{R}\left(\frac{1}{T} - \frac{1}{T^o}\right) + \frac{5}{2}\ln\frac{T^o}{T} \, , \tag{C.8}$$

and thus

$$S = S_o(\rho_p + \rho_{1s})(T/1\mathrm{K})^{-3/2} \exp[158000 K/T] \, .$$

Here we have used the ideal gas law to replace P and $13.6\,\mathrm{eV} = 1.58 \cdot 10^5$ K. S_o is a number depending on the reference state (T^o, P^o), which thermodynamics does not reveal.

However, in Statistical Mechanics we learn that the chemical potential of an ideal system of point-like particles is $\mu_i = RT \ln[\rho_i \Lambda_{T,i}^3] + e_i^{(o)}$,[1] where

$$\Lambda_{T,i} = \sqrt{\frac{2\pi \hbar^2}{m_i k_B T}} \tag{C.9}$$

---

[1] We apply this chemical potential to describe the reference state. The attendant densities ρ_i in this case are all equal to ρ, i.e. they are not the same as the ρ_i introduced at the beginning of this section!.

is the thermal wavelength, k_B is Boltzmann's constant and $e_i^{(o)}$ is an internal contribution to the particle's chemical potential in the above sense. Setting the masses of the proton and the 1s hydrogen equal, i.e. $m_p = m_{1s}$, we obtain with some patience

$$S_o = 4.14 \cdot 10^{-22} \, \text{m}^3 \, . \tag{C.10}$$

Finally, we also need to know $\rho_p + \rho_{1s}$, which is given by

$$\rho_p + \rho_{1s} \approx (1 - Y) \frac{\Omega_{m,b}\rho_{c,0}}{m_p} (T/T_0)^3 \, . \tag{C.11}$$

The quantity $\Omega_{m,b}\rho_{c,0}(T/T_0)^3$ is the baryonic mass density at the time when the radiation temperature is T, as discussed in Sect. 9.5 (cf. (9.23) in conjunction with (9.10) and (9.17), i.e. $\Omega_{m,b} \approx 0.05$ and $\rho_{c,0} \approx 9 \cdot 10^{-27} \, \text{kg m}^{-3}$). The current temperature of the background radiation is $T_0 \approx 2.7$ K. The factor $1 - Y$, where Y is the primordial helium fraction, is roughly $1 - 0.25 = 0.75$. With this we find

$$S \approx 4 \cdot 10^{-24} \, (T/1\text{K})^{3/2} \exp[158000K/T] \, . \tag{C.12}$$

The result is shown in Fig. C.1. We can see that X has essentially dropped to zero at 3000 K.

However, as stated in Weinberg's *Cosmology* [26], the above reasoning gives the right order of magnitude of the temperature of the steep decline in fractional ionisation, but it is not correct in detail. For instance, capture of a free electron into the ground state produces a photon, which in turn has more than sufficient energy to ionise another hydrogen atom. Therefore this process yields no net decrease in ionisation. Details of the full calculation of $X(T)$ including attendant references are given in Weinberg's book ([26] Sect. 2.3). The improved $X(T)$ is less sharp, rising in the range from 2500 to 4500 K, and possess an inflection point at around 3500 K.

Remark We briefly want to motivate two important numbers, i.e. the above primordial helium fraction $Y \approx 0.25$ and the baryon to photon ratio $\eta \sim 10^{-9}$.

When the temperature of the universe was around 10^{10} K, corresponding to about 1 MeV (cf. Table 7.1) the reactions

$$p + \bar{\nu}_e \leftrightarrow n + e^+ \qquad p + e^- \leftrightarrow n + \nu_e$$

established a thermal distribution of protons p and neutrons n. Here ν_e and $\bar{\nu}_e$ are the electron neutrino and its antineutrino, whereas e^- is the electron and e^+ the positron. Thermal equilibrium means that the number ratio n_n/n_p of neutrons and positrons is given by

$$\frac{n_n}{n_p} \approx \exp[-(m_n - m_p)c^2/(k_B T)] \approx e^{-1.5 \cdot 10^{10} K/T} \overset{T=10^{10}\text{K}}{\approx} \frac{1}{5} \tag{C.13}$$

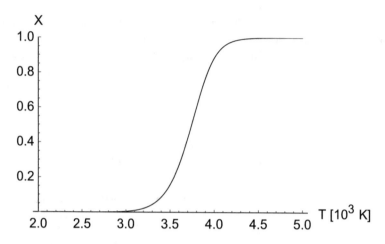

Fig. C.1 X, the fraction of ionised hydrogen, versus temperature, T. Note that $X = (\sqrt{1+4S} - 1)/(2S)$ with S given by (C.12)

The first link in the nucleosynthetic chain, along which helium is produced, is

$$p + n \leftrightarrow d + \gamma , \tag{C.14}$$

where d is deuterium and γ indicates a photon. We can expect that Big Bang nucleosynthesis will take off when

$$\eta^{-1} \exp[-E_B/(k_B T)] \sim 1 , \tag{C.15}$$

where $E_B = 2.2\,\text{MeV}$ is the deuterium binding energy. Essentially the left side in (C.15) is the rate determining factor in the above reaction equation (from right to left), describing the reaction velocity in terms of an Arrhenius law. Inserting $\eta^{-1} \sim 10^9$ from above yields $T \sim 0.1\,\text{MeV} \sim 10^9\,\text{K}$. To good approximation all neutrons now quickly end up in ^4He, whose abundance thus follows via

$$Y = \frac{4 \cdot \frac{n_n}{2}}{1 \cdot n_n + 1 \cdot n_p} = 2\frac{n_n/n_p}{1 + n_n/n_p} . \tag{C.16}$$

Inserting $n_n/n_p \approx 1/5$ form above yields $Y \approx 1/3$. But during the two minutes or so when the temperature decreased from 10^{10} to $10^9\,\text{K}$, the number of neutrons were reduced due to their short half-life of about 14.8 min by a factor $e^{-2/14.8} \approx 0.87$. This in turn reduces the helium abundance to $Y \approx 0.29$—a value in somewhat closer accord with observation (most of the helium abundance today is due to this period!).

Our rough considerations do illustrate the intimate relation of Y and η to the nucleosynthesis in the very early universe and its sensitivity to temperature. Two recommended references, where the interested reader can find detailed information, are [42] and [43]. The latter reference recommends for the abundance of ^{4}He and the baryon to photon ratio

$$\boxed{Y \approx 0.25} \quad \text{and} \quad \boxed{\eta \approx 6 \cdot 10^{-10}} \quad , \tag{C.17}$$

respectively.

Appendix D
Simple Views on Dark Matter Halos

Since the early 1930s astronomical, astrophysical and theoretical observations have strengthened our believe in the existence of a sizeable amount of gravitating matter, which in all other respects does not provide proof or even hints of its existence (see [44]). The first such observations were reported by the Swedish astronomer Knut Lundmark (1930) and the Swiss astronomer Fritz Zwicky (1933). Zwicky in 1933 suggested that dark matter might be the reason for the high velocity dispersion of galaxies in the Coma Cluster. The virial theorem, i.e. $2K = -U$, where K and U are the total kinetic and potential energy of the galaxy cluster, allows to relate the observed velocities of the galaxies to the observed mass in the cluster. However, the latter was found to be considerably less than what was reasonable considering the velocities. Vera Rubin and William K. Ford in the 1970s observed the same unexplainably rapid orbits of stars in individual galaxies (rotation curves). Here the mass M inside a stars orbit of radius r is related to its orbital velocity v via $GM/r^2 \sim v^2/r$ or $v \sim 1/\sqrt{r}$. But the v does not decrease. Instead it becomes constant at large r, which in turn suggested the presence of additional invisible or dark matter. This we shall discuss in some detail below. More evidence came from gravitational lensing experiments. The deflection of light was found to be inconsistent with the observed visible matter (e.g. [45]). A lensing experiment, with an additional twist, was looking at the Bullet Cluster of galaxies. The additional twist are the relative shifts of the centers of gravity of the ordinary matter distributions, compared to that of their dark matter halos according to the lensing effect of the latter. The interpretation is that two galaxy clusters, both being the centers of dark matter halos, did collide in the distant past. The viscous interaction of the ordinary matter (gas clouds) and the lack of it in the case of the dark matter then caused the aforementioned relative shifts. Further evidence for missing matter comes from the Big Bang nucleosynthesis (BBN) (cf. our estimate of the matter energy density based on the baryon to photon ratio). Also, the analysis of the CMB power spectrum as well as the formation of structure in the universe appear to require amounts of dark matter consistent with BBN. Finally, we should not forget that dark matter has a favourable effect on our comparison of

© Springer Nature Switzerland AG 2020
R. Hentschke and C. Hölbling, *A Short Course in General Relativity and Cosmology*, Undergraduate Lecture Notes in Physics,
https://doi.org/10.1007/978-3-030-46384-7

Fig. D.1 Sketch of the
model mass distribution. The
mass element (darker
shading) experiences a
gravitational force due to the
mass $M(h)$, where h is the
radius of the central
spherical region containing
this mass (lighter shading)

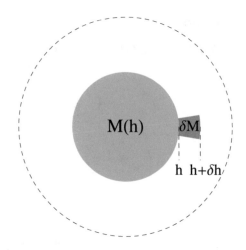

Fig. D.1 Sketch of the model mass distribution. The mass element (darker shading) experiences a gravitational force due to the mass $M(h)$, where h is the radius of the central spherical region containing this mass (lighter shading)

the standard model of cosmology with the data documenting the accelerated cosmic expansion.

For the sake of simplicity the following discussion focusses exclusively on dark matter in the context of rotation curves of galaxies, i.e. the unexpectedly large velocity of stars far from the galactic center (e.g. [46]; even though this does not seem to pertain to all rotation curves [47]). The discussion of dark matter halos is quite complex (e.g. [48] or [49]) and by far exceeds what we are able to discuss here without inflating the topic beyond proportion. We shall begin by assuming the simplest of models, which, even though plagued by serious problems, will provide some valuable insights.

D.1   A Spherical Equilibrium Model

We assume that the dark matter forms a stable spherical halo centred on a galaxy, whose mass we do not consider here unless stated otherwise. The sketch in Fig. D.1 depicts a spherically symmetric region in space exhibiting increased dark matter density. A volume element containing the (dark matter) mass δM is subject to a small radial pressure difference $\delta P(h)$ preventing the collapse of the halo. This pressure difference is given by

$$P(h) - P(h + \delta h) = \frac{Gm\rho(h)M(h)}{h^2}\delta h , \qquad (D.1)$$

where G is the gravitational constant. The quantity m is the mass of a single dark matter particle in the halo and $\rho(h)$ is the particle number density at the radial density h from the center. Thus

$$\frac{dP(h)}{dh} = -\frac{Gm\rho(h)M(h)}{h^2} . \tag{D.2}$$

Note that

$$M(h) = 4\pi m \int_0^h dh' h'^2 \rho(h') . \tag{D.3}$$

In addition we shall assume $P(h) \propto \rho(h)$, i.e.

$$\beta P(h) = \rho(h) . \tag{D.4}$$

where β is a constant.

Putting everything together yields

$$\frac{dy(x)}{dx} = -\frac{y(x)}{x^2} \int_0^x dx' x'^2 y(x') , \tag{D.5}$$

where $y = \rho/\rho_o$ and $x = h/h_o$ with $\rho_o = \rho(0)$ and

$$h_o = (4\pi Gm^2 \beta \rho_o)^{-1/2} . \tag{D.6}$$

Differentiating (D.5) one more time with respect to x yields

$$\frac{1}{x^2} \frac{d}{dx} x^2 \frac{d}{dx} \ln y(x) + y(x) = 0 \tag{D.7}$$

or

$$yy'' - y'^2 + 2\frac{y}{x}y' + y^3 = 0 . \tag{D.8}$$

Here $y' = dy(x)/dx$ and $y'' = d^2y(x)/dx^2$.

One can quickly find a solution of this differential equation by inserting the power law $y = \alpha x^p$. The result is $p = -2$ and $\alpha = 2$ (cf. the discussion of the isothermal sphere in Sect. 4.3 of [48]). This means that the density diverges at the center. However, solutions with finite $y(0)$ and $y'(0) = 0$ do exist as well. A power series solution of the above differential equations in the limit of small x is

$$y(x) = 1 - \frac{x^2}{6} + \frac{x^4}{45} - \frac{61x^6}{22680} + \mathcal{O}(x^8) \tag{D.9}$$

The full numerical solution of (D.8), i.e. $\rho_{IGH}(h)$, obtained under the condition that the density remains finite, is the solid black line in Fig. D.2.

Here the index IGH stands for ideal gas halo, referring to the obvious correspondence of (D.4) to the ideal gas equation of state. We did not emphasize this previously, because the underlying Maxwellian velocity distribution within each volume element

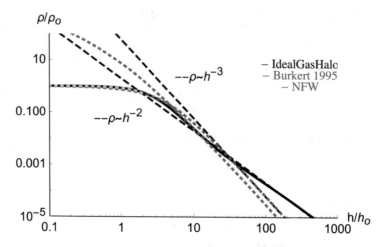

Fig. D.2 Reduced radial dark matter halo density, ρ/ρ_o, versus reduced radial distance from the center, h/h_o. Solid black line: ρ_{IGH} obtained numerically from (D.8); ρ_{IGH} obtained using the expansion (D.9); green line: magenta dashed line: the NFW empirical density profile; red dashed line: the empirical Burkert profile. Thin dashed lines represent the two limiting power laws h^{-2} and h^{-3}

of the isothermal sphere can be obtained for the case of stellar dynamics in galaxies as well. This however is conceptually different from equilibration in a gas (cf. Sect. 4.3 in [48]).

The expansion (D.9) agrees with the numerical solution out to $h/h_o \approx 1$. In the other limit, i.e. large h/h_o, the numerical solution approaches the aforementioned power law. Included in this figure are two frequently applied empirical density distributions, i.e.

$$\rho(h) \propto \frac{1}{(h/h_{\text{NFW}})(1 + h/h_{\text{NFW}})^2} \tag{D.10}$$

used by Navarro, Frenk and White [50] (NFW) and

$$\rho(h) \propto \frac{1}{(h/h_B + 1)((h/h_B)^2 + 1)} \tag{D.11}$$

suggested by Burkert [51]. The quantities h_{NFW} and h_B are adjustable parameters. Note also that the form of the latter expression is in close accord with $\rho_{\text{IGH}}(h)$. Only at large h does the Burkert density profile decreases $\propto h^{-3}$, as does the NFW profile, whereas ρ_{IGH} decreases more slowly and approaches h^{-2}. Obviously, a dark matter density profile which continues as $\propto h^{-2}$ yields a divergent dark matter mass, indicating that our simple model needs serious modification at large distances.[2]

---

[2]Of course, this is also the case if $\rho(h) \propto h^{-3}$.

D.2   Rotation Curves

An immediate plausibility test is the rotation velocity of stars, which is given by

$$v(h)^2 = \frac{GM(h)}{h} . \tag{D.12}$$

In dimensionless form this becomes

$$(v(h)/v_o)^2 = \frac{1}{x} \int_0^x dx' x'^2 y(x') , \tag{D.13}$$

where

$$v_o = (m\beta)^{-1/2} . \tag{D.14}$$

The typical rotation velocity of stars, v_o, is directly related to the thermal velocity of the dark matter particles (i.e. from hereon we assume $\beta = 1/(k_B T)$, where k_B is Boltzmann's constant and T is temperature). This is not really surprising, because both move in the same gravitational field.

The result depicted in Fig. D.3 (top panel) is encouraging. It shows that the distribution $\rho_{\text{IGH}}(h)$ yields the nearly constant velocity desired at large h in contrast to the $h^{-1/2}$-law expected in the absence of dark matter.[3,4] Note that the time for one orbit at large h is a significant fraction of today's age of the universe, e.g. close to 10% at $h = 35$ kpc. Figure D.3 (bottom panel) also includes the integral ratio of luminous to dark matter up to a given distance from the center. Obviously, a dark matter density profile which continues as $\propto h^{-2}$ will lead to a diverging dark matter mass. This is something that needs to be dealt with (cf. above). On the other hand, even this simple model helps to rationalise the large proportion of dark matter in comparison to luminous matter currently favoured in the literature (cf. (9.17)).

Nesti and Salucci [52] derive the mass model of the Milky Way. These authors use the Burkert cored dark matter halo profile, from which they infer $m\rho(0) \approx 2.7 \cdot 10^{-21}$ kg/m^3 (i.e. $4 \cdot 10^7$ M$_\odot$/kpc^3, where M$_\odot \approx 2 \cdot 10^{30}$ kg and 1 kpc $\approx 3.1 \cdot 10^{19}$ m) as well as the radius of the core, which is $h_{\text{core}} \approx 2.8 \cdot 10^{20}$ m (i.e. about 10 kpc). This value for $m\rho(0)$ is somewhat higher than the value $0.53 \cdot 10^{-21}$ kg/m^3 (i.e. 0.3 GeV/cm^3) given in Weinberg's book ([26]; p. 195).[5] However, here we are

---

[3]This law is for orbits outside a spherical mass distribution. If the visible matter in a galaxy forms a 'disk', then the $h^{-1/2}$-law is somewhat modified. Nevertheless, the rotation velocity decreases at large h, in contrast to the observations.

[4]This is seen already if we insert $y \propto x^p$ with $p = -2$ into the integral in (D.13).

[5]We can estimate a similar number just by looking at the experimental data depicted in Fig. D.3 (even though the data are not for our galaxy). The 'velocity plateau' begins at around $h \approx 10$ kpc. This is roughly where the dark matter content of a sphere of radius h can be expected to be equal to its content of ordinary matter. From (D.12) we obtain $M(h \approx 10$ kpc), which we insert into the attendant energy density, i.e. $0.5M(h)c^2/(4\pi h^3/3)$. The result is close to Weinberg's number.

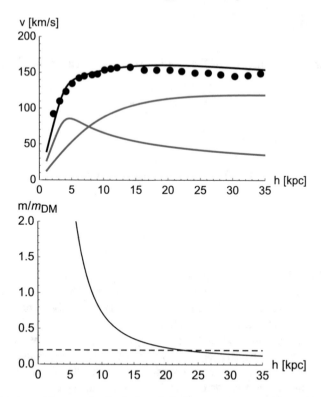

Fig. D.3 Top: A specific example of a measured rotation curve in comparison to this model. Solid symbols: data from Bottema et al. [53]. Black line: fit to the data based on the sum of dark matter, according to (D.13) using the present ideal gas model, and a mock distribution of luminous matter using a power-law-like mass distribution of the form $\propto (1 + (h/h_o)^7)$. Red lines: rotation curves of the two contributions individually. Bottom: Ratio of luminous matter to dark matter integrated to the indicated distance from the galactic center. Black line: result corresponding to the two red lines in the upper panel. Dashed line: value taken from (9.17)

interested in rough estimates and because Nesti and Salucci also provide an estimate for the core radius we shall continue to use their values.

Rotation curves in the literature exhibit different plateau velocities, i.e. rotation velocities far from the galactic center. Here we concentrate on the range between 100 to 300 km/s. From (D.13) follows that the plateau velocity of this model, i.e. the velocity at large h, is $v \approx 1.4 v_o$.[6] We use this relation to estimate v_o based on the data in the top panel of Fig. D.3. The resulting v_o in conjunction with (D.14) yields m/T, which, using (D.6), finally yields h_o.

The rotation velocity about 50 kpc from the core of our galaxy is roughly $v = 250$ km/s [54]. Using this value we find $m/T \approx 0.43 \cdot 10^{-33}$ kg/K, which, with $m\rho(0) \approx 2.7 \cdot 10^{-21}$ kg/m^3 and (D.6), yields

---

[6] Actually, this is true if luminous matter is neglected.

Table D.1 Model results

v (km/s)	$10^{33}\, m/T$ (kg/K)	m/T (keV/K)	$10^{22}\, m\rho_o$ (kg/m^3)
100	2.7	1.5	4.3
150	1.2	0.7	9.7
200	0.7	0.4	17
250	0.4	0.24	27
300	0.3	0.2	39

$$h_o = 1.2 \cdot 10^{20} \text{ m} = 3.8 \text{ kpc} . \tag{D.15}$$

Note that $h_{\text{core}}/h_o = 2.4$, in very good accord with the extend of the flat part of the ideal gas density profile in Fig. D.2.

Table D.1 compiles the values for m/T and $m\rho_o$ for different values of the rotation velocity v far from the galactic center, where it is dominated by dark matter. The first column lists the rotation velocities. The second and third column list the attendant values of m/T. Apparently the model yields an increase of dark matter temperature with increasing rotation velocity. This is due to the higher core density, which follows with the above value for h_o from (D.6). The relation between core density and temperature will be discussed below. It is worth noting that $mc^2 \gg k_B T$.

D.3 The Virial Approach

Let's explore the halo from a slightly different angle by applying the virial theorem, i.e.

$$2K = -U \tag{D.16}$$

in the case of the gravitational potential. Here K is its kinetic energy, whereas U is the potential energy of the halo. The potential energy of the halo is given by

$$U = -G \int_o^{M_c} \frac{M(h)}{h} dM , \tag{D.17}$$

i.e. the halo of total mass $M_c = M(h_c)$ is build by adding shells of mass dM to an existing mass $M(h)$ given in terms of the density by (D.3). Thus

$$U = -(4\pi m)^2 G \int_o^{h_c} dh \rho(h) h \int_0^h dh' \rho(h') h'^2 . \tag{D.18}$$

In Fig. D.2 we have included the power law

$$\frac{\rho(h)}{\rho_o} = \frac{\alpha}{(h/h_o)^2} , \tag{D.19}$$

where $\alpha = 2$ (cf. above), which approaches ρ_{IGH} at large h/h_o. If we use this approximation for all h instead of ρ_{IGH}, the potential energy is simply

$$U \approx -(4\pi m\alpha\rho_o h_o^2)^2 G h_c . \tag{D.20}$$

Analogously the kinetic energy can be worked out:

$$K = \frac{3}{2}k_B T M_c/m \approx 6\pi k_B T c\rho_o h_o^2 h_c . \tag{D.21}$$

The final result, according to the Virial theorem in (D.16), is given by

$$m/T \approx \frac{3}{4\pi\alpha}\frac{k_B}{m\rho_o G h_o^2} \overset{(D.6)}{=} \underbrace{\frac{3}{\alpha}}_{\approx 1} m/T . \tag{D.22}$$

The right side is obtained after insertion of the previous equation for h_o. Note that the ill-defined parameter h_c has disappeared. Note also that the two approaches to the problem, i.e. (D.1) and (D.16), are of course intimately related through the application of the ideal gas law to the pressure in the first case and to the kinetic energy in the second. Thus, (D.22) is a check of overall consistency rather than a new result. However, there is one new piece of information.

The total energy of our isothermal halo according to the virial theorem is given by

$$E = \frac{1}{2}U = -K . \tag{D.23}$$

Aside from the problem with its actual size we also find that the heat capacity $C_V = \partial E/\partial T|_V$ is negative, which means that it is unstable from a statistical mechanical point of view. However, application of statistical mechanics to self-gravitating systems is far from straightforward (cf. Chap. 7 in [48] for more detail). This is essentially is due to the infinite range of the interactions. We already ignored this difficulty when we started to use the ideal gas law by setting $\beta = 1/(k_B T)$. In statistical mechanics the ideal gas law results from the leading term in a cluster expansion of the free energy, i.e. which means that there are no interactions between particles. This in turn requires that inter particle interactions possess a finite range.

Let us compute t_c and t_{pres} according to (10.47) and (10.48) for the plateau part of a dark matter halo. Using $\rho_m = 2.7 \cdot 10^{-21}$ kg/m^3 we obtain $t_c \sim 2 \cdot 10^{15}$ s. Next we use $R = h_{\mathrm{core}} = 2.8 \cdot 10^{20}$ m and $c_s = \sqrt{\delta P/\delta\rho_m} = \sqrt{k_B T/m}$. Inserting $m/T \sim$

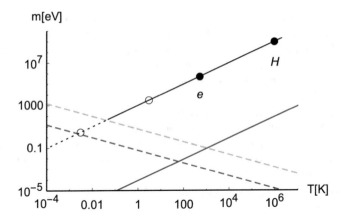

Fig. D.4 Black lines: Particle mass, m, versus temperature, T, according to (D.24). Here e and H indicate particles with the mass of an electron and a hydrogen atom, respectively. The open symbols indicate particles whose mass corresponds to $T = 3\,K$ and $T = 0.003\,K$, respectively. The dashed line indicates the regime in which the classical approach breaks down. Coloured lines: Green and red lines correspond to the lower of the two equations in (D.28), i.e. the lines indicate BE condensation at the respective density. The green line is obtained with $\rho(0)$ (halo center) and the red line is obtained for $10^{-5} \cdot \rho(0)$. The blue line corresponds to $mc^2 = k_B T$

$10^{33}\,kg/K$ we find $t_{\mathrm{pres}} \sim 3 \cdot 10^{15}$ s. Our estimates are rather crude and thus we can only conclude that both time scales are similar.

D.4 Problems with the Classical Picture: Fermions Versus Bosons

At this point we want to explore possible bounds on the mass of dark matter particles imposed by the model—despite its difficulties. According to Table D.1 a fairly typical value for m/T is

$$m/T \approx 1\ \mathrm{keV/K} \tag{D.24}$$

Figure D.4 depicts (D.24) together with a number of examples.[7] For instance, if a halo would consist of hydrogen, then the gas is a hot plasma. At a distance of about 10 kpc from the galactic center it would contain between 10^3 to 10^4 (ionized) hydrogens per m^3, which is detectable (cf. [55]; p. 68).

Thus far our discussion has focussed on today's dark matter halos. However, it has been shown, mostly via simulation work, that the current structures formed by

---

[7]In order for us to get a feeling for this relation, let us calculate m/T for the case of air at sea level and $T = 300\,K$. The result is $m/T \sim 10^5$ keV/K.

baryonic matter require that dark matter was cold from almost the very beginning of the universe. Cold means non-relativistic dark matter, because otherwise it would have washed out the lumpy structure created by inflation (e.g. [27]). However, we do not have to go into the details here to make the following points. Either dark matter was created cold or it must have decoupled very early from the other components in the universe. The latter means that if it was a hot gas, then it would have adiabatically cooled much faster in comparison to radiation, i.e. $T_r \sim a^{-1}$ but $T_{DM} \sim a^{-2}$. Thus, we may even expect that the halo temperature, if temperature is applicable, is the result of the virial theorem, i.e. potential energy is redistributed and increases the kinetic energy of the halo gas as described above. At this point we are back on the black line in Fig. D.4.

Note that low temperatures bring about a problem with the classical point of view, because the latter requires

$$\lambda_T \ll \rho^{-1/3} , \tag{D.25}$$

where $\lambda_T = \hbar(2\pi\beta/m)^{1/2}$ is the thermal wavelength. We illustrate this using two example temperatures. For instance, a temperature 1000 times less than the temperature of the microwave background, i.e. $T = 3/1000 \, \mathrm{K}$, according to (D.24) yields $m \approx 5 \cdot 10^{-36} \, \mathrm{kg}$ and thus $\lambda_T \approx 6 \cdot 10^{-4} \, \mathrm{m}$, whereas $\rho(0)^{-1/3} \approx 10^{-5} \, \mathrm{m}$. This is outside the classical domain. If on the other hand $T = 3 \, \mathrm{K}$ we find $m \approx 5 \cdot 10^{-33} \, \mathrm{kg}$ and thus $\lambda_T \approx 6 \cdot 10^{-7} \, \mathrm{m}$. Because $\rho(0)^{-1/3} \approx 10^{-4} \, \mathrm{m}$, this system is inside the classical domain. The cross-over occurs around $T \approx 0.05 \, \mathrm{K}$ and $m \approx 50 \, \mathrm{eV}$. When the temperature is less, then strong deviations from the linear relation between pressure and density (cf. (D.4)) are expected, in which case it is not clear how to obtain the radial density distribution needed to explain the flat rotation curves.

In the case of Fermions without direct interaction the pressure approaches $P \approx (2\epsilon_F/5)\rho$ when $T \to 0$, where $\epsilon_F = (2m)^{-1}\hbar^2(6\pi^2\rho)^{2/3}$ is the Fermi energy (e.g. [56]). The relation between pressure and density for Fermions in this limit, i.e. $\lambda_T \gg \rho^{-1/3}$, is $P \propto \rho^{5/3}$. Using this instead of (D.4) leads to a differential equation analogous to (D.5), which, if we insert again $y \propto x^p$, yields $p = -10$! This means that the radial density decrease is much too fast. In particular one obtains $v(x) \propto x^{-4}$ from (D.13) instead of $v(x) = const$.

This also makes another possible dark matter candidate unlikely—the (ordinary) neutrino. Neutrinos are Fermions. Recent estimates state that the neutrino mass is less than about $2 \, \mathrm{eV}$ [43]. In order for this mass to fit in our theory there must be a corresponding halo temperature which is roughly $T \approx 10^{-4} \, \mathrm{K}$ according to (D.24). This in turn yields $\lambda_T \approx 83 \cdot 10^{-5} \, \mathrm{m}$ and $\rho(0)^{-1/3} \approx 1.1 \cdot 10^{-5} \, \mathrm{m}$. The overall neutrino mass limit from cosmological observations is less than $0.2 \, \mathrm{eV}$ [57], which yields $\lambda_T \approx 8.3 \cdot 10^{-3} \, \mathrm{m}$ and $\rho(0)^{-1/3} \approx 5.1 \cdot 10^{-6} \, \mathrm{m}$. This is deep inside the non-classical domain, which, as we have just discussed, does not yield reasonable rotation velocities. On the other hand, above we had argued that we expect that today neutrinos have a temperature close to the photon background temperature. But this argument was based on the assumption that they have no mass or behave relativistically, i.e.

$mc^2 \ll k_B T$. For the above masses and a neutrino temperature on the order of 3 K this is not satisfied. In any case, at present neutrinos do not seem to be viable candidates.

In the case of Bosons without direct interaction the particles undergo a transition to a condensate at $\lambda_T^3 \rho \approx 2.6$.[8] But how would the transition be approached. Consider the following scaling argument:

$$\lambda_T \rho^{1/3} \sim T^{-1/2} a^{-1} \sim (a^{-\theta})^{-1/2} a^{-1} \sim a^{\theta/2-1} \ . \tag{D.26}$$

If the particles behave as non-relativistic particles (not in equilibrium with radiation) then $\theta = 2$, i.e. the transition effectively is not approached—one way or the other.[9] But for $\theta = 1$ (equilibrium with radiation) we find $\lambda_T \rho^{1/3} \sim a^{-1/2}$. Here the transition is possible. But it would be a transition into the classical domain, because a is growing, rather than the opposite. However, it also is possible that a condensate is formed during formation of the halo itself.

The situation is illustrated by the cartoon in Fig. D.5. The dark matter particles forming a halo are in equilibrium with the dark matter outside the halo, i.e. the chemical potential of the particles in the halo, μ_h, is equal to the chemical potential of the particles outside the halo, μ. The chemical potential μ is zero.[10] The dark matter particles are somewhere on the thick line in the diagram. Note that the density of Bosons in the ground state along this line is given by

$$\rho_o = \rho_c \left(1 - T/T_c\right)^{3/2} \qquad (T \leq T_c) \tag{D.27}$$

The quantities ρ_c and T_c are the the dark matter critical particle number density and the critical temperature outside the halo at the lower end of the thick line in the diagram in Fig. D.5. If momentarily we focus on the particles which are not in the ground state (even though $\mu = 0$, then for these particles

$$\beta P V = \frac{V}{\lambda_T^3} \zeta(5/2)$$

$$\text{and} \tag{D.28}$$

$$\lambda_T^3 \rho = \zeta(3/2) \approx 2.61 \ .$$

The second equation is plotted in Fig. D.4 for two different densities corresponding the core and to the distant fringes of a dark matter halo. These equations are valid along the entire thick line in the graph shown in Fig. D.5. We can combine the equations (D.28) to yield $P \approx 0.513 \, k_B T \rho$. But note also that according to (D.27), which really follows from the second equation (D.28), we have $T = T_c (\rho/\rho_c)^{2/3}$ and thus $P \propto \rho^{5/3}$. Above we had looked at the same relation in the case of Fermions, which does not lead to halo mass density profiles in accord with experimental rotation curves.

---

[8]Here the Boson spin is zero. But the general picture does not change if it is not.

[9]In principle the condensate may have existed from the very start.

[10]Strictly speaking $\mu = \epsilon_o$, where ϵ_o is the one-particle ground state energy.

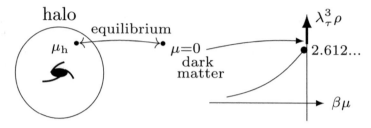

Fig. D.5 Dark matter particles in the halo in equilibrium with the dark matter outside the halo. All dark matter is somewhere on the thick line in the sketch of the reduced density $\lambda_T \rho$ versus the reduced chemical potential, $\beta\mu$

Can there possibly be an effect due to the gravitational interaction? The effect of a homogeneous gravitational field $\vec{g}$ on the BE condensation of a Boson gas in a box of height L is explained in some detail in for instance Chap. 6 of [58]. In the case of T_c one finds

$$T_c(\kappa) = T_c(0) \left(1 + \frac{8\sqrt{\pi}}{9\zeta(3/2)} \sqrt{\kappa} \right) \tag{D.29}$$

($\kappa = \beta_c(0)mgL$) in the limit of small κ. Let's estimate κ. We assume $g(h_o) = GM(h_o)/h_o^2$, where h_o is given by (D.15) and $M(h_o)$ is calculated via (D.3) using (D.19). Here we assume that the particle mass m is 10^{-3} eV (cf. the discussion at the end of this section). The result is $g(h_o) \approx 5 \cdot 10^{-10}$ m/s². We further assume $L = 500h_o$. This is of course quite arbitrarily based on Fig. D.2. The attendant density is $\rho(L) \approx \rho(0)10^{-5}$. This yields $T_c(0) = 2\pi\hbar^2/(mk_B)(\rho(L)/2.6)^{2/3} \approx 900$ K. For κ follows $\kappa \approx 10^{-5}$. Thus, at first glance, the effect of gravitation in the halo on the particles outside the ground state is quite negligible.

But how about the particles in the ground state? This has already been studied in the paper by Böhmer and Harko [59]. The approach is based on the Gross–Pitaevskii equation, which describes the ground state solution of a N-particle Bose system in an external field V_{ext}. Here V_{ext} is the field due to the gravitation interaction between the particles and obeys Poisson's equation. Both equations may be combined to yield a non-linear homogeneous differential equation for the effective ground state wave function ψ. The number density inside the dark matter halo is then given by $|\psi|^2$. Böhmer and Harko employ an solution schema based on the Thomas–Fermi approximation. In the end they find solutions whose radial profile yields flat rotation curves. But we are not convinced. In addition, others claim that the Thomas–Fermi approximation is not appropriate in the present case [60]. Apparently the better solutions yield rotations curves which do not fit the experiments. Nevertheless, the relation between dark matter halos and BE condensation is an active field of research (cf. the review by Suárez, Robles and Matos [61]).

One current candidate for dark matter, which is in accord with the above discussion, is the axion. This hypothetical particle is a boson with spin zero, in the mass

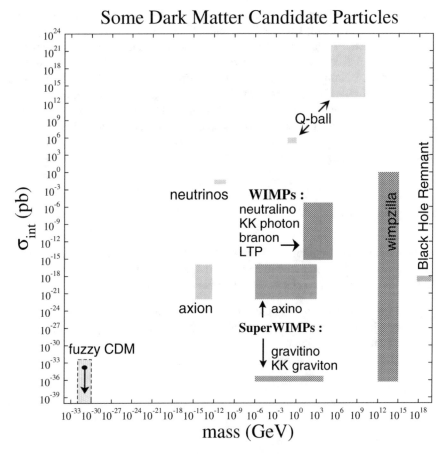

Fig. D.6 Dark matter particle candidates according to a 2007 report of the Dark Matter Scientific Assessment Group of the US National Science Foundation (https://www.nsf.gov/mps/ast/aaac/ dark_matter_scientific_assessment_group/dmsag_final_report.pdf). The figure is a copy of Fig. 20 of the report

range roughly 10^{-5} to 10^{-3} eV or $1.8 \cdot 10^{-41}$ to $1.8 \cdot 10^{-39}$ kg. It originated as an (initially unrecognised) side effect of trying to fix a completely different problem in particle physics.[11] Assuming that relation (D.24) is valid, the corresponding temper-

---

[11] The strong nuclear force, which is described in the standard model of particle physics by quantum chromodynamics (QCD), is invariant under a symmetry transformation that encompasses both spatial reflection (parity P) and charge conjugation (C). There is no fundamental reason for the absence of CP symmetry breaking terms in the strong interaction and in fact in the weak interaction they are present. Peccei and Quinn suggested in 1977 that the absence of symmetry breaking terms might be explained dynamically with the help of a scalar field. Weinberg and Wilczek later showed that this implies the existence of a scalar particle, the axion.

atures, thermal wavelength and number densities (in the halo center) are 10^{-8} K and 10^{-6} K, 1.7 m and 170 m, $1.5 \cdot 10^{20}$ m^{-3} and $1.5 \cdot 10^{18}$ m^{-3}.

Pretty much the whole range of dark matter candidates under investigation is compiled in Fig. D.6, taken from the 2007 report of the Dark Matter Scientific Assessment Group of the US National Science Foundation. Not all of them fit into the our above halo model of course—especially the heavy ones. The search for dark matter is still intensifying. New theoretical ideas are produced every day and older ones are improved or extended. The same happens on the experimental side. All large experiments in particle physics are involved and new experiments are set up or planned continuously. This truly is an exciting time.

D.5 Problems

1. Apply the virial theorem to a homogeneous dust cloud and determine its radius after virialization compared to its initial radius.
2. DM and DE effects on classical tests of general relativity:

   (a) Does the presence of Dark Matter alter our result for the perihelion precession of Mercury—one of the classical tests of general relativity theory? Provide an order of magnitude estimate.
   (b) How about Dark Energy?

Appendix E
Solutions to the Problems

E.1 Chapter 2

1. Let us view this problem from the rest frame of the first person and let the coordinate origin ($x = 0, t = 0$) be located at the event of the meeting (cf. the spacetime diagram). Note that the traffic light is located at $x = -240$ m at $t = 0$.

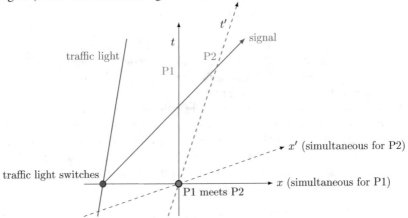

In this coordinate system the second person moves with a velocity $v = 0.3$ m/s $\simeq 10^{-9}c$ into the positive x-direction. For the first person, the switching event happens simultaneously to the meeting event, i.e. at $t = 0$. For the second person, it thus happens at

© Springer Nature Switzerland AG 2020
R. Hentschke and C. Hölbling, *A Short Course in General Relativity
and Cosmology*, Undergraduate Lecture Notes in Physics,
https://doi.org/10.1007/978-3-030-46384-7

$$t' = \gamma\left(t - \frac{v}{c}\frac{x}{c}\right)$$

$$\simeq \frac{1}{\sqrt{1 - 10^{-18}}}\left(10^{-9}\frac{240\,\text{m}}{c}\right)$$

$$\simeq 24 \cdot 10^{-8}\,\text{m}\frac{1}{3 \cdot 10^8\,\text{m/s}}$$

$$= 8 \cdot 10^{-16}\,\text{s}\,,$$

which is an of course unnoticeable bit later than the meeting.

The Andromeda galaxy has a distance of about 2.5 million light-years or ~2.4 · 10^{22} m from Earth. If we substitute this distance instead of the distance to the traffic light into the above calculation, we obtain

$$t' = \gamma\left(t - \frac{v}{c}\frac{x}{c}\right)$$

$$\simeq \frac{1}{\sqrt{1 - 10^{-18}}}\left(10^{-9}\frac{2.4 \cdot 10^{22}\text{m}}{c}\right)$$

$$\simeq 24 \cdot 10^{12}\text{m}\frac{1}{3 \cdot 10^8\,\text{m/s}}$$

$$= 8 \cdot 10^4\,\text{s}\,,$$

which is slightly less than a day. We thus see that even if two observers move at nonrelativistic speed, they agree on simultaneity only in a limited spatial region. Or, in other words, Galilei transformations are the low velocity limit of Lorentz transformations only in a small region.

2. Let us first look at the situation from the point of view of the garage. We take the coordinate origin ($x = 0$, $t = 0$) of the garage's as the event when the rear end of the car passes the rear door (which then closes instantaneously). We take the same event as the coordinate origin of the cars rest frame ($x' = 0$, $t' = 0$). In the cars rest frame, the car has of course a length of 4 m, i.e. $x' = 4$ m is the front end of the car. From the Lorentz transformation we generically have

$$x' = \gamma(x - vt)\,.$$

So when the back door closes (at $t = 0$) we have

$$x = \frac{x'}{\gamma} = \sqrt{1 - \frac{1}{4}}4\,\text{m} = 2\sqrt{3}\,\text{m} \simeq 3.46\,\text{m}\,,$$

which is less than the length of the garage. Thus, from the point of view of the garage, the car does fit inside.

Now let us take the perspective of the car. Viewed from within the car, the back door closes at $t' = 0$. Where is the front door ($x = 4$ m) at that time? We have

$$t' = \gamma\left(t - \frac{v}{c}\frac{x}{c}\right)$$

and thus the time t in the garage's rest frame for which $t' = 0$ at the front door is

$$t = \frac{v}{c}\frac{x}{c} = \frac{1}{2}\frac{4\text{ m}}{c} = \frac{2\text{ m}}{c}.$$

In the car's reference frame, the front door at that time is at a coordinate

$$x' = \gamma(x - vt)$$
$$= \frac{1}{\sqrt{1 - \frac{1}{4}}}\left(4\text{ m} - \frac{c}{2}\frac{2\text{ m}}{c}\right)$$
$$= \frac{2}{\sqrt{3}}3\text{ m} = 2\sqrt{3}\text{ m} \simeq 3.46\text{ m}.$$

From the point of view of the car, the garage has contracted and the car never fits inside.

3. (a) For Alice, the spinning disk simply has a circumference of $C_A = 2\pi r$. For Bob however the situation is different. Since Bob moves relative to Alice with a velocity $v = \omega r$, Alice observes that the measuring rods used by Bob are shortened by a factor $1/\gamma = \sqrt{1 - v^2/c^2}$ in the direction of motion. Since the circumference moves tangential to the disc, all measuring rods along the circumference shrink accordingly. If Bob thus puts a tape measure around the circumference of the disc, Alice will see the markings shrunk by a factor $1/\gamma$ and thus it will give a circumference that is larger by a factor γ, i.e.

$$C_B = \gamma C_A = \frac{1}{\sqrt{1 - \left(\frac{\omega r}{c}\right)^2}}2\pi r.$$

Note that Alice and Bob agree on the radius, since it is perpendicular to the motion of the disc. Bob thus lives in a world where the circumference of a circle is larger than $2\pi r$ and thus his geometry is not Euclidean (it is called hyperbolic as we will learn later). This is of course caused by the acceleration of his frame of reference, which is not inertial.

(b) Let us put Alice's coordinate origin ($\vec{x} = 0$) into the center of the disk and the rotation axis be the z-axis. Let us also set $t = 0$ in this coordinate system so that Bob follows a trajectory

$$\vec{x}(t) = \begin{pmatrix} r\cos\omega t \\ r\sin\omega t \\ 0 \end{pmatrix}.$$

For this trajectory we have

$$dx = r\omega \sin \omega t dt \qquad dy = -r\omega \cos \omega t dt \qquad dz = 0$$

and thus

$$d\tau = \sqrt{dt^2 - \tfrac{1}{c^2}dx^2 - \tfrac{1}{c^2}dy^2 - \tfrac{1}{c^2}dz^2}$$
$$= \sqrt{1 - \tfrac{r^2\omega^2}{c^2}}\, dt$$
$$= dt/\gamma \, .$$

Bob experiences the proper time τ along his path, while Alice experiences her coordinate time t. For him a shorter time

$$T_B = \frac{T}{\gamma} = \sqrt{1 - \frac{r^2\omega^2}{c^2}}\, T$$

passes until he encounters Alice again. This is commonly known as the twin paradox.

4. The gravitational potential energy of a point mass m at a distance r from a point mass M is

$$V = -G\frac{Mm}{r} \, .$$

The kinetic energy of the mass m is

$$T = \frac{mv^2}{2}$$

and thus the total energy is

$$E = \frac{mv^2}{2} - G\frac{Mm}{r} \, .$$

For the system not to be gravitationally bound we require $E \geqslant 0$. In the limiting case $E = 0$ we have

$$\frac{mv_e^2}{2} = G\frac{Mm}{r} \, ,$$

where v_e now is the escape velocity. Solving for it we find

$$v_e = \sqrt{\frac{2GM}{r}} \, .$$

Now let us assume $v_e = c$. We obtain

$$r_s = \frac{2GM}{c^2} \, .$$

If we insert in the mass of the Sun $M_\odot \simeq 1.99 \cdot 10^{30}$ kg, we find

$$
\begin{aligned}
r_{s\odot} &= \frac{2GM_\odot}{c^2} \\
&\simeq \frac{2 \cdot 6.67 \cdot 10^{-11} \cdot 1.99 \cdot 10^{30} \text{ kg m}^3 \text{s}^2}{9 \cdot 10^{16} \text{m}^2 \text{ kg s}^2} \\
&\simeq 2.95 \text{ km} \ .
\end{aligned}
$$

5. (a) We have the differentials

$$
\begin{aligned}
dt &= d\rho \sinh \alpha + \rho \cosh \alpha \, d\alpha \\
dx &= d\rho \cosh \alpha + \rho \sinh \alpha \, d\alpha
\end{aligned}
$$

and therefore

$$
\begin{aligned}
d\tau^2 &= dt^2 - dx^2 \\
&= \rho^2(\cosh^2 \alpha - \sinh^2 \alpha)d\alpha^2 + d\rho^2(\sinh^2 \alpha - \cosh^2 \alpha) \ . \\
&= \rho^2 d\alpha^2 - d\rho^2
\end{aligned}
$$

The resulting metric is

$$
g_{\mu\nu} = \begin{pmatrix} \rho^2 & 0 \\ 0 & -1 \end{pmatrix} \ .
$$

(b) For $\rho = \text{const}$ we simply have $d\tau = \pm \rho d\alpha$, i.e.

$$
a^\mu = \frac{d^2 x^\mu}{d\tau^2} = \frac{d^2 x^\mu}{\rho^2 d\alpha^2} = \frac{1}{\rho}\begin{pmatrix} \sinh \alpha \\ \cosh \alpha \end{pmatrix} \ .
$$

The norm of the acceleration is

$$
a_\mu a^\mu = \frac{1}{\rho^2}(\sinh^2 \alpha - \cosh^2 \alpha) = -\frac{1}{\rho^2} \ .
$$

(c) We compute the Christoffel symbols according to

$$
\Gamma^\mu_{\lambda\sigma} = \frac{1}{2}g^{\mu\nu}(\partial_\sigma g_{\lambda\nu} + \partial_\lambda g_{\nu\sigma} - \partial_\nu g_{\lambda\sigma}) \ .
$$

The inverse metric is easy to find

$$
g^{\mu\nu} = \begin{pmatrix} \rho^{-2} & 0 \\ 0 & -1 \end{pmatrix} \ .
$$

The only non-trivial derivative of a metric component is

$$
\partial_\rho g_{\alpha\alpha} = 2\rho
$$

and thus

$$\Gamma^{\rho}_{\alpha\alpha} = \rho \qquad \Gamma^{\alpha}_{\alpha\rho} = \Gamma^{\alpha}_{\rho\alpha} = \frac{1}{\rho}$$

with all other components vanishing. The equations of motion thus read

$$\frac{d^2\alpha}{d\tau^2} + \frac{2}{\rho}\frac{d\rho}{d\tau}\frac{d\alpha}{d\tau} = 0 \quad \text{and} \quad \frac{d^2\rho}{d\tau^2} + \rho\frac{d\alpha}{d\tau}\frac{d\alpha}{d\tau} = 0 \, .$$

E.2 Chapter 3

1. (a) The equation for the invariant line element may be rewritten as

$$g_{\mu\nu}\frac{dx^{\mu}}{d\tau}\frac{dx^{\nu}}{d\tau} = 1 \, .$$

Multiplying the covariant force

$$F^{\mu} = m\frac{d}{d\tau}\frac{dx^{\mu}}{d\tau}$$

with $g_{\mu\nu}dx^{\nu}/d\tau$ we find

$$\begin{aligned}
F_{\mu}\frac{dx^{\mu}}{d\tau} &= m\left(\frac{d}{d\tau}\frac{dx^{\mu}}{d\tau}\right)g_{\mu\nu}\frac{dx^{\nu}}{d\tau} \\
&= m\frac{d}{d\tau}\left(g_{\mu\nu}\frac{dx^{\mu}}{d\tau}\frac{dx^{\nu}}{d\tau}\right) - m\frac{dx^{\mu}}{d\tau}\frac{d}{d\tau}g_{\mu\nu}\frac{dx^{\nu}}{d\tau} \\
&= -\frac{dx^{\mu}}{d\tau}mg_{\mu\nu}\frac{d}{d\tau}\frac{dx^{\nu}}{d\tau} \\
&= -\frac{dx^{\mu}}{d\tau}F_{\mu} \, ,
\end{aligned}$$

which implies

$$F_{\mu}\frac{dx^{\mu}}{d\tau} = 0 \, .$$

(b) The covariant version of the potential gradient equation is

$$F_{\mu} = m\partial_{\mu}\varphi \, .$$

Multiplication with $dx^{\mu}/d\tau$ yields

$$m \frac{dx^\mu}{d\tau} \partial_\mu \varphi = F_\mu \frac{dx^\mu}{d\tau} = 0 \ .$$

The left hand side of this equation is a total derivative, i.e.

$$m \frac{d\varphi}{d\tau} = 0 \ .$$

Since m is constant this implies that the potential φ is constant along an arbitrary path.

2. (a) We first compute the total differentials

$$dx = \sin\theta \cos\varphi \, dr - r \sin\theta \sin\varphi \, d\varphi$$
$$dy = \sin\theta \sin\varphi \, dr + r \sin\theta \cos\varphi \, d\varphi$$
$$dz = \cos\theta \, dr$$

from which we obtain

$$\begin{aligned}
ds^2 &= dx^2 + dy^2 + dz^2 \\
&= \sin^2\theta((\cos\varphi dr - r \sin\varphi d\varphi)^2 \\
&\quad + (\sin\varphi dr + r \cos\varphi d\varphi)^2) + \cos^2\theta dr^2 \\
&= \sin^2\theta(dr^2 + r^2 d\varphi^2) + \cos^2\theta dr^2 \\
&= dr^2 + r^2 \sin^2\theta \, d\varphi^2
\end{aligned}$$

so that the metric is

$$g_{ab} = \begin{pmatrix} 1 & 0 \\ 0 & r^2 \sin^2\theta \end{pmatrix} \ .$$

We could at this point define a new angular coordinate $\phi = \varphi \sin\theta$, which rescales φ by a constant factor. In these coordinates the metric would be

$$\hat{g}_{ab} = \begin{pmatrix} 1 & 0 \\ 0 & r^2 \end{pmatrix} \ .$$

The inverse metric is given by

$$g^{ab} = \begin{pmatrix} 1 & 0 \\ 0 & r^{-2} \sin^{-2}\theta \end{pmatrix} \qquad \hat{g}^{ab} = \begin{pmatrix} 1 & 0 \\ 0 & r^{-2} \end{pmatrix} \ .$$

The only nontrivial coordinate derivative of the metric is

$$g_{\varphi\varphi,r} = 2r \sin^2\theta \qquad \hat{g}_{\phi\phi,r} = 2r \ .$$

Thus the only nonzero Christoffel symbols are easily found as

$$\Gamma^{\varphi}_{\varphi r} = \Gamma^{\varphi}_{r\varphi} = \frac{1}{r} \qquad \Gamma^{r}_{\varphi\varphi} = -r \sin^2 \theta$$

$$\hat{\Gamma}^{\phi}_{\phi r} = \hat{\Gamma}^{\phi}_{r\phi} = \frac{1}{r} \qquad \hat{\Gamma}^{r}_{\phi\phi} = -r \, .$$

(b) The coordinate transformation is

$$X = r \cos \phi = r \cos(\varphi \sin \theta)$$
$$Y = r \sin \phi = r \sin(\varphi \sin \theta) \, ,$$

which implies

$$
\begin{aligned}
dX &= dr \cos \phi - r \sin \phi d\phi \\
&= dr \cos(\varphi \sin \theta) - r \sin(\varphi \sin \theta) \sin \theta d\varphi \\
dY &= dr \sin \phi + r \cos \phi d\phi \\
&= dr \sin(\varphi \sin \theta) + r \cos(\varphi \sin \theta) \sin \theta d\varphi
\end{aligned}
$$

and thus

$$dX^2 + dY^2 = dr^2 + r^2 d\phi^2 = dr^2 + r^2 \sin^2 \theta d\varphi^2 = ds^2 \, .$$

(c) We compute the covector y_a by multiplying the contravariant vector y^a with the metric

$$y_a = g_{ab} y^b = \begin{pmatrix} 1 \\ 0 \end{pmatrix} \, .$$

In this coordinate notation the vector is constant, thus its normal derivatives vanish $y_{a,b} = 0$. We thus have

$$y_{a;b} = -\Gamma^c_{ab} y_c = -\Gamma^r_{ab} \, .$$

There is only one non-vanishing Christoffel symbol $\Gamma^r_{\varphi\varphi}$ resp. $\hat{\Gamma}^r_{\phi\phi}$ that will contribute, thus

$$y_{\varphi;\varphi} = r \sin^2 \theta \qquad y_{\phi;\phi} = r$$

with all other covariant derivatives vanishing. The difference $y_{a;b} - y_{b;a}$ thus vanishes in all cases.

We now repeat this exercise in the cartesian coordinates X^a. We know generically that a contravariant vector transforms as

$$X^a = \frac{\partial X^a}{\partial y^b} y^b \, .$$

Inserting the partial derivatives we have computed above results in

$$
\begin{aligned}
X^X &= \tfrac{\partial X}{\partial r} y^r + \tfrac{\partial X}{\partial \varphi} y^\varphi \\
&= \cos(\varphi \sin \theta) \\
&= \cos \phi \\
&= \tfrac{X}{r} \\
&= \tfrac{X}{\sqrt{X^2+Y^2}}
\end{aligned}
$$

$$
\begin{aligned}
X^Y &= \tfrac{\partial Y}{\partial r} y^r + \tfrac{\partial Y}{\partial \varphi} y^\varphi \\
&= \sin(\varphi \sin \theta) \\
&= \sin \phi \\
&= \tfrac{Y}{r} \\
&= \tfrac{Y}{\sqrt{X^2+Y^2}}
\end{aligned}
$$

and thus the vector, in 2-D cartesian coordinates, reads

$$
X^a = \frac{1}{\sqrt{X^2 + Y^2}} \begin{pmatrix} X \\ Y \end{pmatrix} .
$$

The covariant vector has the same components, since the metric is just $g_{ab} = \delta_{ab}$, i.e.

$$
X_a = \frac{1}{\sqrt{X^2 + Y^2}} \begin{pmatrix} X \\ Y \end{pmatrix} .
$$

All derivatives of the metric vanish in cartesian coordinates, thus all Christoffel symbols vanish, too. The covariant derivative is simply the ordinary derivative and we get

$$
X_{X;X} = \frac{\sqrt{X^2+Y^2} - \frac{X^2}{\sqrt{X^2+Y^2}}}{X^2+Y^2} = \frac{Y^2}{(X^2+Y^2)^{3/2}}
$$

$$
X_{X;Y} = -\frac{XY}{(X^2+Y^2)^{3/2}}
$$

$$
X_{Y;X} = -\frac{XY}{(X^2+Y^2)^{3/2}}
$$

$$
X_{Y;Y} = \frac{\sqrt{X^2+Y^2} - \frac{Y^2}{\sqrt{X^2+Y^2}}}{X^2+Y^2} = \frac{X^2}{(X^2+Y^2)^{3/2}}
$$

which implies $X_{a;b} - X_{b;a} = 0$. Alternatively this result may be obtained directly by transforming $y_{a;b}$ into the new coordinates. We have

$$X_{a;b} = \frac{\partial y^c}{\partial X^a} \frac{\partial y^d}{\partial X^b} y_{c;d}$$
$$= \frac{\partial \varphi}{\partial X^a} \frac{\partial \varphi}{\partial X^b} y_{\varphi;\varphi}$$
$$= \frac{\partial \varphi}{\partial X^a} \frac{\partial \varphi}{\partial X^b} r \sin^2 \theta$$
$$= \frac{\partial \varphi}{\partial X^a} \frac{\partial \varphi}{\partial X^b} \sqrt{X^2 + Y^2} \sin^2 \theta .$$

Since we have

$$\varphi = \sin^{-1} \theta \, \text{atan} \, \frac{Y}{X} ,$$

the partial derivatives are

$$\frac{\partial \varphi}{\partial X} = -\frac{Y}{X^2 + Y^2} \frac{1}{\sin \theta} \qquad \frac{\partial \varphi}{\partial X} = \frac{X}{X^2 + Y^2} \frac{1}{\sin \theta} ,$$

i.e.

$$X_{a;b} = \begin{pmatrix} \frac{Y^2}{(X^2+Y^2)^{3/2}} & -\frac{XY}{(X^2+Y^2)^{3/2}} \\ -\frac{XY}{(X^2+Y^2)^{3/2}} & \frac{X^2}{(X^2+Y^2)^{3/2}} \end{pmatrix} ,$$

which we also obtained by direct differentiation.

(d) First we compute the covariant vector

$$\xi_a = g_{ab}\xi^b = \begin{pmatrix} 0 \\ r^2 \sin^2 \theta \end{pmatrix} .$$

The covariant derivative of this vector is

$$\xi_{a;b} = \delta_{a\varphi}\delta_{br} 2r \sin^2 \theta - \Gamma^\varphi_{ab} r^2 \sin^2 \theta$$
$$= \delta_{a\varphi}\delta_{br} 2r \sin^2 \theta - (\delta_{a\varphi}\delta_{br} + \delta_{b\varphi}\delta_{ar}) r \sin^2 \theta ,$$
$$= (\delta_{a\varphi}\delta_{br} - \delta_{b\varphi}\delta_{ar}) r \sin^2 \theta$$

i.e. $\xi_{a;b} + \xi_{b;a} = 0$ thus ξ^a is a Killing vector.

3. (a) We have the differentials

$$dx = r \cos \theta \cos \varphi \, d\theta - r \sin \theta \sin \varphi \, d\varphi$$
$$dy = r \cos \theta \sin \varphi \, d\theta + r \sin \theta \cos \varphi \, d\varphi$$
$$dz = -r \sin \theta \, d\theta$$

and therefore

$$ds^2 = dx^2 + dy^2 + dz^2 = r^2 d\theta^2 + r^2 \sin^2 \theta d\varphi^2 \ ,$$

i.e. the resulting metric is

$$g_{ab} = \begin{pmatrix} r^2 & 0 \\ 0 & r^2 \sin^2 \theta \end{pmatrix}$$

and the inverse metric

$$g^{ab} = \begin{pmatrix} r^{-2} & 0 \\ 0 & r^{-2} \sin^{-2} \theta \end{pmatrix} \ .$$

The only nonzero derivative of the metric with respect to a coordinate is

$$g_{\varphi\varphi,\theta} = 2r^2 \sin \theta \cos \theta$$

and thus the nonzero Christoffel symbols are

$$\Gamma^{\varphi}_{\theta\varphi} = \Gamma^{\varphi}_{\varphi\theta} = \frac{\cos \theta}{\sin \theta} \qquad \Gamma^{\theta}_{\varphi\varphi} = - \sin \theta \cos \theta \ .$$

(b) The covariant vector is

$$y_a = \begin{pmatrix} 0 \\ r \sin \theta \end{pmatrix} \ .$$

Therefore $y^a y_a = 1$, i.e. y^a has unit norm. The covariant derivatives are

$$\begin{aligned} y_{\theta;\theta} &= -\Gamma^c_{\theta\theta} y_c \\ &= 0 \end{aligned}$$

$$\begin{aligned} y_{\theta;\varphi} &= -\Gamma^c_{\theta\varphi} y_c \\ &= -r \cos \theta \end{aligned}$$

$$\begin{aligned} y_{\varphi;\theta} &= y_{\varphi,\theta} - \Gamma^c_{\varphi\theta} y_c \\ &= r \cos \theta - r \cos \theta \\ &= 0 \end{aligned} \qquad ,$$

$$\begin{aligned} y_{\varphi;\varphi} &= -\Gamma^c_{\varphi\varphi} y_c \\ &= 0 \end{aligned}$$

i.e. the difference

$$y_{\varphi;\theta} - y_{\theta;\varphi} = r \cos \theta$$

is not zero.

4. The Lie derivative of a covariant vector is

$$\pounds_\xi A_\mu = \xi^\alpha \partial_\alpha A_\mu + A_\alpha \partial_\mu \xi^\alpha .$$

Using (3.45) and (3.49) we can rewrite this as

$$\begin{aligned}
\pounds_\xi A_\mu &= \xi^\alpha (\nabla_\alpha A_\mu + \Gamma^\beta_{\alpha\mu} A_\beta) + A_\alpha (\nabla_\mu \xi^\alpha - \Gamma^\alpha_{\mu\beta} \xi^\beta) \\
&= \xi^\alpha \nabla_\alpha A_\mu + \Gamma^\beta_{\alpha\mu} A_\beta \xi^\alpha + A_\alpha \nabla_\mu \xi^\alpha - \Gamma^\alpha_{\mu\beta} A_\alpha \xi^\beta , \\
&= \xi^\alpha \nabla_\alpha A_\mu + A_\alpha \nabla_\mu \xi^\alpha
\end{aligned}$$

which is a covariant vector. Similarly we find for the Lie derivative of a contravariant vector

$$\begin{aligned}
\pounds_\xi A^\mu &= \xi^\alpha \partial_\alpha A^\mu - A^\alpha \partial_\alpha \xi^\mu \\
&= \xi^\alpha \nabla_\alpha A^\mu - A^\alpha \nabla_\alpha \xi^\mu ,
\end{aligned}$$

i.e. this is a contravariant vector. For the Lie derivative of the metric tensor we have

$$\begin{aligned}
\pounds_\xi g_{\mu\nu} &= \xi^\alpha \partial_\alpha g_{\mu\nu} + g_{\alpha\nu} \partial_\mu \xi^\alpha + g_{\mu\beta} \partial_\nu \xi^\beta \\
&= \xi^\alpha \nabla_\alpha g_{\mu\nu} + g_{\alpha\nu} \nabla_\mu \xi^\alpha + g_{\mu\beta} \nabla_\nu \xi^\beta .
\end{aligned}$$

The first term vanishes because the covariant derivative of the metric is zero, so we can write the Killing equation as

$$\pounds_\xi g_{\mu\nu} = \nabla_\mu \xi_\nu + \nabla_\nu \xi_\mu = 0 .$$

5. Under transformation from coordinates x^μ to coordinates y^μ the differential space-time volume element transforms according to

$$d^4 y = \left| \frac{\partial y}{\partial x} \right| d^4 x .$$

Here $|\partial y / \partial x|$ is the Jacobian of the coordinate transformation, i.e. the determinant of the matrix whose components are $\partial y^\mu / \partial x^\nu$. The Jacobian can be expressed in terms of the determinants of the metrics. We see this via

$$g_{\alpha\beta}(y) = g_{\mu\nu}(x) \frac{\partial x^\mu}{\partial y^\alpha} \frac{\partial x^\nu}{\partial y^\beta} .$$

Taking the determinant yields

$$\det g(y) = \det g(x) \left| \frac{\partial x}{\partial y} \right|^2$$

or

$$\sqrt{-\det g(y)}\, \mathrm{d}^4 y = \sqrt{-\det g(x)}\, \mathrm{d}^4 x .$$

The minus sign is inserted, because physical spacetimes have a negative determinant.

E.3   Chapter 4

1. The sketches below illustrate the different quantities. Note the circumference of the green circle is $2\pi R$. The circumference of the cut-and-flattened cone is $2\pi l$. Thus

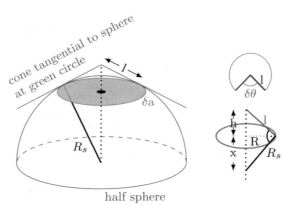

half sphere

$$2\pi R = 2\pi l - \delta\theta\, l .$$

With the help of the drawing on the right we find $l^2/R^2 = R^2(R_s^2 - R^2)^{-1} + 1 \approx 1 + (R/R_s)^2$ ($R_s \gg R$). Thus

$$1 - \frac{l}{R} \approx \frac{1}{2}\frac{\delta a}{\pi R_s^2} .$$

The combination of the two equations yields the desired result.
2. This relation is most easily established in locally flat coordinates. We have

$$R_{\mu\nu\alpha\beta;\gamma} + R_{\mu\nu\beta\gamma;\alpha} + R_{\mu\nu\gamma\alpha;\beta}$$

$$= \frac{1}{2}(g_{\mu\beta,\nu\alpha\gamma} + g_{\nu\alpha,\mu\beta\gamma} - g_{\nu\beta,\mu\alpha\gamma} - g_{\mu\alpha,\nu\beta\gamma})$$

$$+ \frac{1}{2}(g_{\mu\gamma,\nu\beta\alpha} + g_{\nu\beta,\mu\gamma\alpha} - g_{\nu\gamma,\mu\beta\alpha} - g_{\mu\beta,\nu\gamma\alpha})$$

$$+ \frac{1}{2}(g_{\mu\alpha,\nu\gamma\beta} + g_{\nu\gamma,\mu\alpha\beta} - g_{\nu\alpha,\mu\gamma\beta} - g_{\mu\gamma,\nu\alpha\beta})$$

$$= 0 .$$

3. (a) In two dimension, all indices are restricted to run from 0 to 1 only. Because of the antisymmetry property, the first two and the last two indices must differ for the component to be nonzero, which just leaves us with the four nonzero components

$$\mathcal{R}_{0101} \qquad \mathcal{R}_{0110} \qquad \mathcal{R}_{1001} \qquad \mathcal{R}_{1010}$$

that are related as follows

$$\mathcal{R}_{0101} = -\mathcal{R}_{0110} = -\mathcal{R}_{1001} = \mathcal{R}_{1010} .$$

The symmetry property gives no additional restriction and the cyclic property

$$0 = \mathcal{R}_{0101} + \mathcal{R}_{0011} + \mathcal{R}_{0110} = \mathcal{R}_{0101} - \mathcal{R}_{0101}$$

is also trivially fulfilled. The curvature scalar is found to be

$$\mathcal{R} = g^{AC} g^{BD} \mathcal{R}_{ABCD}$$

$$= g^{00} g^{11} \mathcal{R}_{0101} + g^{01} g^{10} \mathcal{R}_{0110} + g^{10} g^{01} \mathcal{R}_{1001} + g^{11} g^{00} \mathcal{R}_{1010}$$

$$= 2(g^{00} g^{11} - g^{10} g^{01}) \mathcal{R}_{0101}$$

$$= 2 \det(g)^{-1} \mathcal{R}_{0101} .$$

Using the notation $g = \det g$ for the metric determinant, we can write

$$\mathcal{R}_{0101} = -\mathcal{R}_{0110} = -\mathcal{R}_{1001} = \mathcal{R}_{1010} = \frac{g}{2}\mathcal{R} .$$

(b) In $3 + 1$ dimensions we simply expand $\varepsilon^{\mu\nu\alpha\beta} R_{\mu\nu\alpha\beta}$ taking into account the other symmetries of the Riemann tensor

$$0 = \varepsilon^{\mu\nu\alpha\beta} R_{\mu\nu\alpha\beta}$$

$$= 8R_{\mu\nu\alpha\beta} - 8R_{\beta\mu\nu\alpha} - 8R_{\nu\alpha\beta\mu}$$

$$= 8R_{\mu\nu\alpha\beta} + 8R_{\mu\beta\nu\alpha} + 8R_{\mu\beta\nu\alpha} ,$$

which is the desired cyclicity relation.

In $2 + 1$ dimensions, at least one index occurs twice. If the duplicate index appears within the last three, we have

$$R_{ijjk} + R_{ijkj} + R_{ikjj} = R_{ijjk} - R_{iijjk} = 0$$

by antisymmetry in the last index alone. In the other case where the first index is shared with one of the last 3, we have

$$R_{iijk} + R_{ijki} + R_{ikij} = R_{ijki} + R_{ijik} = 0 \, ,$$

again implied by the other symmetry properties, so the cyclic permutation symmetry is no additional restriction.

(c) In D dimensions, the Riemann tensor has D^4 components. Antisymmetry in the first and last two components reduces the number of independent components to $(D(D-1)/2)^2$. The exchange symmetry between first and last two indices reduces this further to

$$\frac{1}{2} \frac{D(D-1)}{2} \left(\frac{D(D-1)}{2} + 1 \right) = \frac{1}{8}(D^2 - D)(D^2 - D + 2)$$

or 21 components for $D = 4$. As we have seen in the previous problem, the cyclicity just imposes one extra condition in four dimensions, so we have in total 20 independent components of the Riemann tensor.

For $D = 3$ we have 6 independent components and no additional restriction from cyclicity.

4. A vanishing energy momentum tensor implies that the Ricci tensor vanishes. The Ricci tensor is symmetric, thus it has 10 independent components in $3 + 1$ dimensions. In the previous exercise we determined that the Riemann tensor has 20 independent components, so there are 10 components of the Riemann tensor that are not fixed by a zero Ricci tensor, which allows nontrivial vacuum solutions. In $1 + 1$ dimensions the Riemann tensor is fully determined even by the Ricci scalar, so there is no nontrivial vacuum solution.

In $2 + 1$ dimensions, the Ricci tensor has 6 independent components as does the Riemann tensor. It therefore has also no nontrivial vacuum solutions.

5. In a previous exercise we have found the nonzero Christoffel symbols for a sphere of radius r in spherical coordinates to be

$$\Gamma^\varphi_{\theta\varphi} = \Gamma^\varphi_{\varphi\theta} = \frac{\cos\theta}{\sin\theta} \qquad \Gamma^\theta_{\varphi\varphi} = -\sin\theta\cos\theta \, .$$

Thus the nonzero derivatives are

$$\Gamma^\varphi_{\theta\varphi,\theta} = \Gamma^\varphi_{\varphi\theta,\theta} = 2 - \frac{1}{\sin^2\theta} \qquad \Gamma^\theta_{\varphi\varphi,\theta} = 2\sin^2\theta - 1 \, .$$

The Riemann tensor generically is given by

$$R^A_{BCD} = \Gamma^A_{DB,C} - \Gamma^A_{CB,D} + \Gamma^A_{CE}\Gamma^E_{BD} - \Gamma^A_{DE}\Gamma^E_{BC} \ .$$

We can lower the first index by multiplying with the metric

$$g_{AF} = \begin{pmatrix} r^2 & 0 \\ 0 & r^2\sin^2\theta \end{pmatrix} \ .$$

Computing one nonzero component, e.g.

$$\begin{aligned}
R_{\theta\varphi\theta\varphi} &= g_{\theta\theta}R^\theta_{\varphi\theta\varphi} \\
&= r^2(\Gamma^\theta_{\varphi\varphi,\theta} - \Gamma^\theta_{\theta\varphi,\varphi} + \Gamma^\theta_{\theta E}\Gamma^E_{\varphi\varphi} - \Gamma^\theta_{\varphi E}\Gamma^E_{\varphi\theta}) \\
&= r^2(2\sin^2\theta - 1 + \cos^2\theta) \\
&= r^2\sin^2\theta
\end{aligned}$$

provides us with the curvature scalar

$$\mathcal{R} = \frac{2}{g}R_{\theta\varphi\theta\varphi} = \frac{2}{r^4\sin^2\theta}r^2\sin^2\theta = \frac{2}{r^2} \ .$$

The nonzero components of the Ricci tensor are also easily found as

$$\mathcal{R}_{\varphi\varphi} = g^{\theta\theta}\mathcal{R}_{\theta\varphi\theta\varphi} = \sin^2\theta \qquad \mathcal{R}_{\theta\theta} = g^{\varphi\varphi}\mathcal{R}_{\varphi\theta\varphi\theta} = 1$$

or, perhaps more intuitively

$$\mathcal{R}^\varphi_\varphi = \mathcal{R}^\theta_\theta = \frac{1}{r^2} \ .$$

6. (a) The metric and its inverse are

$$g_{\mu\nu} = \begin{pmatrix} 1 & 0 & 0 & 0 \\ 0 & -a^2 & 0 & 0 \\ 0 & 0 & -a^2 & 0 \\ 0 & 0 & 0 & -a^2 \end{pmatrix}$$

$$g^{\mu\nu} = \begin{pmatrix} 1 & 0 & 0 & 0 \\ 0 & -a^{-2} & 0 & 0 \\ 0 & 0 & -a^{-2} & 0 \\ 0 & 0 & 0 & -a^{-2} \end{pmatrix} \ .$$

The only nontrivial derivatives are therefore

$$g_{ii,0} = -2\dot{a}a \ ,$$

which implies that the only nonzero Christoffel symbols are

$$\Gamma^0_{ii} = \dot{a}a \qquad \Gamma^i_{i0} = \Gamma^i_{0i} = \frac{\dot{a}}{a}\,.$$

(b) We start by performing derivatives of the Christoffel symbols. The only nonzero derivatives are

$$\Gamma^0_{ii,0} = \ddot{a}a + \dot{a}^2 \qquad \Gamma^i_{i0,0} = \Gamma^i_{0i,0} = \frac{\ddot{a}a - \dot{a}^2}{a^2}\,.$$

Using the general relation

$$R^\alpha_{\beta\gamma\delta} = \Gamma^\alpha_{\delta\beta,\gamma} - \Gamma^\alpha_{\gamma\beta,\delta} + \Gamma^\alpha_{\gamma\varepsilon}\Gamma^\varepsilon_{\beta\delta} - \Gamma^\alpha_{\delta\varepsilon}\Gamma^\varepsilon_{\beta\gamma}\,.$$

We immediately find

$$R^0_{i0i} = \ddot{a}a + \dot{a}^2 - \dot{a}^2 = \ddot{a}a$$

and thus

$$R_{0i0i} = R_{i0i0} = -R_{i00i} = -R_{0ii0} = \ddot{a}a\,,$$

which fixes 12 elements. The derivative terms do not contribute to any other component, but the product terms allow one other nonzero combination, namely

$$R^i_{jij} = \Gamma^i_{i0}\Gamma^0_{jj} = \dot{a}^2$$

and thus

$$R_{ijij} = -R_{ijji} = -\dot{a}^2a^2$$

fixing another 12 components.

(c) Clearly the Ricci tensor allows for only two independent components, namely

$$\mathcal{R}_{00} = \sum_i \mathcal{R}^i_{0i0} = \sum_i g^{ii}\mathcal{R}_{i0i0} = -3\frac{\ddot{a}}{a}$$

and

$$\mathcal{R}_{jj} = \mathcal{R}^0_{j0j} + \sum_{i \neq j} \mathcal{R}^i_{jij} = \ddot{a}a + 2\dot{a}^2\,.$$

The curvature follows directly

$$R = \sum_{\mu} g^{\mu\mu} R_{\mu\mu} = -3\frac{\ddot{a}}{a} - \frac{3}{a^2}(\ddot{a}a + 2\dot{a}^2) = -6\left(\frac{\ddot{a}}{a} + \frac{\dot{a}^2}{a^2}\right) .$$

(d) The Einstein tensor is

$$\mathcal{G}_{\mu\nu} = \mathcal{R}_{\mu\nu} - \frac{1}{2}g_{\mu\nu}\mathcal{R} = \begin{pmatrix} 3\frac{\dot{a}^2}{a^2} & 0 & 0 & 0 \\ 0 & -2\ddot{a}a - \dot{a}^2 & 0 & 0 \\ 0 & 0 & -2aa - \dot{a}^2 & 0 \\ 0 & 0 & 0 & -\ddot{a}a - 2\dot{a}^2 \end{pmatrix},$$

i.e. we obtain an energy momentum tensor

$$T_{\mu\nu} = \frac{1}{8\pi G} \begin{pmatrix} 3\frac{\dot{a}^2}{a^2} & 0 & 0 & 0 \\ 0 & -2\ddot{a}a - \dot{a}^2 & 0 & 0 \\ 0 & 0 & -2\ddot{a}a - \dot{a}^2 & 0 \\ 0 & 0 & 0 & -\ddot{a}a - 2\dot{a}^2 \end{pmatrix},$$

which for a constant scale factor implies

$$T_{\mu\nu} = 0 .$$

7. We begin with

$$0 = \frac{\partial}{\partial x^\beta} T^{\alpha\beta} = \frac{\partial}{\partial x^\beta} \left[(\rho + P)u^\alpha u^\beta - P\eta^{\alpha\beta} \right] .$$

For $\alpha = 0$ and $\rho + P \approx \rho$ we find

$$0 = \frac{\partial}{\partial x^\beta} T^{0\beta} \approx \frac{\partial}{\partial x^\beta} \left[\rho u^\beta - P\eta^{0\beta} \right]$$

$$= \frac{\partial}{\partial t}[\rho - P] + \frac{\partial}{\partial x}\rho v_x + \frac{\partial}{\partial y}\rho v_y + \frac{\partial}{\partial z}\rho v_z ,$$

i.e. in this limit

$$0 = \frac{\partial}{\partial t}\rho + \vec{\nabla} \cdot \vec{j} .$$

Here $\rho - P \approx \rho$ and $\vec{j} = \rho\vec{v}$ is the mass flux density.
Similarly for $\alpha = 1$ we obtain

$$0 \approx \frac{\partial}{\partial t}\rho v_x + \frac{\partial}{\partial x}\left[\rho v_x^2 + P\right] + \frac{\partial}{\partial y}\rho v_x v_y + \frac{\partial}{\partial z}\rho v_x v_z .$$

Application of $dv_x/dt = \partial v_x/\partial t + (\vec{v} \cdot \vec{\nabla}) v_x$ in conjunction with the continuity equation derived for $\alpha = 0$ yields the desired result.

E.4 Chapter 5

1. First we calculate

$$a(1 - e^2) \approx 86.2 \, \text{AU} \ .$$

For the Sun we had

$$GM_\odot \approx 1.475 \, \text{km} \approx 10^{-8} \, \text{AU} \ ,$$

so for Sagittarius A* we obtain

$$GM \approx 4 \times 10^{-2} \, \text{AU}$$

and therefore the periastron advance per revolution is

$$\delta\phi \approx 6\pi \frac{GM}{a(1 - e^2)} \approx 0.0087 \approx 22' \ .$$

To obtain the periastron advance per century, we need the orbital period. According to Kepler's third law it is

$$T = \sqrt{\frac{4\pi}{GM} a^3} \approx 1680 \, \text{AU} \sqrt{\frac{4\pi}{0.04} 1680} \approx 1.22 \times 10^6 \, \text{AU} \ .$$

To convert this to time units we are familiar with, we use the fact that light from the Sun reaches Earth in $\sim 8.3 \, \text{min}$, so $1 \, \text{AU} \approx 8.3 \, \text{min}$ and therefore

$$T \approx 19.3 \, a$$

and the periastron advance per century is

$$\delta\phi \approx 1°52' \ .$$

2. (a) We can most easily depict this situation graphically:

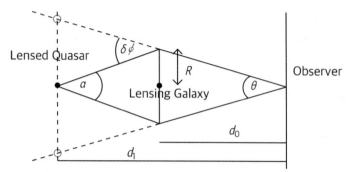

We see that the light deflection is

$$\delta\phi = \frac{\theta + \alpha}{2} \,.$$

From the triangles we obtain

$$d_0\frac{\theta}{2} \approx d_0 \sin\frac{\theta}{2} = R = (d_1 - d_0)\sin\frac{\alpha}{2} \approx (d_1 - d_0)\frac{\alpha}{2} \,.$$

So

$$\alpha \approx \frac{d_0}{d_1 - d_0}\theta$$

and thus the deflection angle must be

$$\delta\phi = \frac{\theta + \alpha}{2} \approx \left(1 + \frac{d_0}{d_1 - d_0}\right)\frac{\theta}{2} = \frac{d_1}{d_1 - d_0}\frac{\theta}{2} \,.$$

We know that the deflection angle is

$$\delta\phi = \frac{4GM}{R} \,,$$

so that finally we can compute the mass

$$M = \delta\phi\frac{R}{4G} = \frac{d_1}{d_1 - d_0}\frac{\theta\, d_0\theta}{2\,8G} = \frac{d_0 d_1}{d_1 - d_0}\frac{\theta^2}{16G} \approx 0.10\frac{\text{lt yr}}{G} \,.$$

The gravitational constant can be expressed as

$$G \approx 6.674 \times 10^{-11}\frac{\text{m}^3}{\text{kg s}^2} \approx 1.56 \times 10^{-13}\frac{\text{lt yr}}{M_\odot}$$

and so the mass of the intermediate galaxy expressed in solar masses is

$$M \approx 6.7 \times 10^{11} M_{\odot} \ .$$

(b) For the unsymmetric case we have two deflection angles $\delta\phi_1$ and $\delta\phi_2$ that add up to

$$\delta\phi_1 + \delta\phi_2 = \theta + \alpha \approx \frac{d_1}{d_1 - d_0}\theta \ .$$

The deflection angles are

$$\delta\phi_i = \frac{4GM}{R_i} \ ,$$

i.e.

$$\frac{d_1}{d_1 - d_0}\theta = 4GM\left(\frac{1}{R_1} + \frac{1}{R_2}\right) \ .$$

We further have

$$\frac{R_1}{R_2} = \frac{\frac{\theta}{2} + \varphi}{\frac{\theta}{2} - \varphi}$$

and thus

$$\frac{d_1}{d_1 - d_0}\theta = 4GM\frac{1}{R_1}\left(1 + \frac{\frac{\theta}{2} + \varphi}{\frac{\theta}{2} - \varphi}\right) \ .$$

Solving for R_1 we obtain

$$R_1 = \left(1 - \frac{d_0}{d_1}\right)\frac{4GM}{\theta}\left(1 + \frac{\frac{\theta}{2} + \varphi}{\frac{\theta}{2} - \varphi}\right) \approx 34708 \, \mathrm{lt} \, \mathrm{yr}$$

and similarly

$$R_2 = \left(1 - \frac{d_0}{d_1}\right)\frac{4GM}{\theta}\left(1 + \frac{\frac{\theta}{2} - \varphi}{\frac{\theta}{2} + \varphi}\right) \approx 20094 \, \mathrm{lt} \, \mathrm{yr} \ .$$

The length difference of the two paths is

$$\Delta = \sqrt{d_0^2 + R_1^2} + \sqrt{(d_1 - d_0)^2 + R_1^2}$$
$$- \sqrt{d_0^2 + R_2^2} - \sqrt{(d_1 - d_0)^2 + R_2^2} \approx 0.26 \, \mathrm{lt} \, \mathrm{yr} \ ,$$

i.e. it takes about 94 days to see the brightness change along the longer path.

3. (a) The nonzero partial derivatives of the metric (omitting elements related by the metric symmetry) are

$$g_{xx,t} = -g_{xx,z} = -g_{yy,t} = g_{yy,z} = -2h'_+ \qquad g_{xy,t} = -g_{xy,z} = -2h'_\times$$

and the inverse metric is given by

$$g^{\mu\nu} = \begin{pmatrix} 1 & 0 & 0 & 0 \\ 0 & -1 + 2h_+ + 4(h_+^2 + h_\times^2) & 2h_\times & 0 \\ 0 & 2h_\times & -1 - 2h_+ + 4(h_+^2 + h_\times^2) & 0 \\ 0 & 0 & 0 & -1 \end{pmatrix}.$$

The nonzero Christoffel symbols are then

$$\begin{aligned}
\Gamma^t_{xx} &= \Gamma^z_{xx} = -\Gamma^t_{yy} = -\Gamma^z_{yy} = h'_+ \\
\Gamma^t_{xy} &= \Gamma^z_{xy} = h'_\times \\
\Gamma^x_{xt} &= -\Gamma^x_{xz} = h'_+ - 2h_+ h'_+ - 2h_\times h'_\times \\
\Gamma^x_{yt} &= -\Gamma^x_{yz} = h'_\times - 2h_+ h'_\times + 2h_\times h'_+ \\
\Gamma^y_{xt} &= -\Gamma^y_{xz} = h'_\times + 2h_+ h'_\times - 2h_\times h'_+ \\
\Gamma^y_{yt} &= -\Gamma^y_{yz} = -h'_+ - 2h_+ h'_+ - 2h_\times h'_\times
\end{aligned}$$

(b) The Ricci tensor is generically given by

$$R_{\mu\nu} = \Gamma^\alpha_{\mu\nu,\alpha} - \Gamma^\alpha_{\alpha\mu,\nu} + \Gamma^\alpha_{\alpha\beta}\Gamma^\beta_{\mu\nu} - \Gamma^\alpha_{\beta\nu}\Gamma^\beta_{\alpha\mu}.$$

To first order in h the quadratic terms vanish. The first term vanishes to all orders because $\Gamma^t_{\mu\nu} = \Gamma^z_{\mu\nu}$ and $h_{,t} = -h_{,z}$ with all other derivatives vanishing. And finally, the second term vanishes to first order because $\Gamma^x_{xt} + \Gamma^y_{yt} = -(\Gamma^x_{xz} + \Gamma^y_{yz}) = -4h_+ h'_+ - 4h_\times h'_\times$ and all other combinations are zero.

(c) We have to evaluate the quadratic terms. To second order the third term vanishes because to first order $\Gamma^x_{x\beta} = -\Gamma^y_{y\beta}$ and $\Gamma^t_{t\beta} = \Gamma^z_{z\beta} = 0$. For the fourth term, all combinations with either $\alpha = t$ or $\beta = t$ are cancelled by the corresponding $\alpha = z$ or $\beta = z$ contributions because there is a sign flip when the index is downstairs but not when it is upstairs. This leaves only $\alpha, \beta \in \{x, y\}$ and the nonvanishing components to second order are

$$\Gamma^\alpha_{\beta t}\Gamma^\beta_{\alpha t} = \Gamma^\alpha_{\beta z}\Gamma^\beta_{\alpha z} = -\Gamma^\alpha_{\beta t}\Gamma^\beta_{\alpha z} = -\Gamma^\alpha_{\beta z}\Gamma^\beta_{\alpha t} = 2h'^2_+ + 2h'^2_\times.$$

Together with the nonvanishing components from the second term

$$\begin{aligned}
\Gamma^\alpha_{\alpha t,t} &= \Gamma^\alpha_{\alpha z,z} - \Gamma^\alpha_{\alpha t,z} = -\Gamma^\alpha_{\alpha z,t} \\
&= -4(h_+ h'_+ + h_\times h'_\times)' = -4(h_+ h''_+ + h'^2_+ + h_\times h''_\times + h'^2_\times).
\end{aligned}$$

This yields

$$R_{\mu\nu} = 2(2h_+h''_+ + h'^2_+ + 2h_\times h''_\times + h'^2_\times) \begin{pmatrix} 1 & 0 & 0 & -1 \\ 0 & 0 & 0 & 0 \\ 0 & 0 & 0 & 0 \\ -1 & 0 & 0 & 1 \end{pmatrix}.$$

The Ricci scalar vanishes, so $G_{\mu\nu} = R_{\mu\nu}$ and using the Einstein equation we find

$$T_{\mu\nu} = \frac{2h_+h''_+ + h'^2_+ + 2h_\times h''_\times + h'^2_\times}{4\pi G} \begin{pmatrix} 1 & 0 & 0 & -1 \\ 0 & 0 & 0 & 0 \\ 0 & 0 & 0 & 0 \\ -1 & 0 & 0 & 1 \end{pmatrix}.$$

(d) To obtain an estimate for the energy radiated off, we integrate the energy density over a spherical shell with radius $R \sim 400\,\mathrm{Mpc} \sim 1.3 \times 10^9\,\mathrm{yr} \sim 4.1 \times 10^{16}$ s. The thickness of the shell is $d \sim 0.1$ s, so the volume of the spherical shell containing the gravitational wave signal is $V \sim 2.1 \times 10^{33}$ s^3. The rms strain in the shell is $h \sim 5 \times 10^{-22}$ and the frequency about $\omega \sim 600$ Hz, so $h^2_{,t} \sim h_{,tt}h \sim \omega^2 h^2 \sim 9 \times 10^{-38}$ s^{-2}. For the energy density we therefore estimate $T^{00} \sim \frac{3}{4\pi G} 9 \times 10^{-38}$ s$^{-2} \sim \frac{1}{G} 2 \times 10^{-38}$ s^{-2} and thus for the total energy we obtain $E \sim G^{-1} 2 \times 10^{-5}$ s. Newtons constant is $G^{-1} \sim 4 \times 10^{35}\,\frac{\mathrm{kg}}{\mathrm{s}} \sim 2 \times 10^5\,M_\odot/s$, so we estimate the total energy radiated off to be equivalent to about 4 solar masses.

4. We use

$$h_{\alpha\beta}(y) = \frac{\partial x^\mu}{\partial y^\alpha} \frac{\partial x^\nu}{\partial y^\beta} h_{\mu\nu}(x) ,$$

i.e. $y^1 + iy^2 = (x^1 + ix^2)e^{-i\phi}$ or $x^1 + ix^2 = y^1 \cos\phi - y^2 \sin\phi + i(y^1 \sin\phi + y^2 \cos\phi)$. Thus

$$\frac{\partial x^1}{\partial y^1} = \cos\phi \quad \frac{\partial x^2}{\partial y^1} = \sin\phi$$
$$\frac{\partial x^1}{\partial y^2} = -\sin\phi \quad \frac{\partial x^2}{\partial y^2} = \cos\phi .$$

We obtain for $h_{11}(y)$ (here $h_{11} = h_+$ and $h_{12} = h_\times$)

$$
\begin{aligned}
h_{11}(y) \quad &= \quad \frac{\partial x^\mu}{\partial y^1} \frac{\partial x^\nu}{\partial y^1} h_{\mu\nu}(x) \\
&= \quad \cos^2\phi\, h_{11}(x) + \sin^2\phi\, h_{22}(x) \\
&\quad + \cos\phi \sin\phi\, h_{12} + \sin\phi \cos\phi\, h_{21}(x) \\
&\overset{h_{22}=-h_{11}, h_{12}=-h_{21}}{=} h_{11}(x)\cos(2\phi) + h_{12}\sin(2\phi)
\end{aligned}
$$

Analogously we find

$$h_{22}(y) = -h_{11}(x)\,\cos(2\phi) - h_{12}\,\sin(2\phi)$$
$$h_{21}(y) = -h_{11}(x)\,\sin(2\phi) + h_{12}\,\cos(2\phi)$$
$$h_{12}(y) = -h_{11}(x)\,\sin(2\phi) + h_{12}\,\cos(2\phi)\ .$$

Thus

$$h_{11}(y) + ih_{12}(y) = h_{11}(x)\,(\cos(2\phi) - i\,\sin(2\phi))$$
$$+ h_{12}(x)\,(\sin(2\phi) + i\,\cos(2\phi))$$

or

$$h_{11}(y) + ih_{12}(y) = (h_{11}(x) + ih_{12}(x))\,e^{-i2\phi}\ .$$

E.5   Chapter 6

1. (a) The only nontrivial equations of motion are those for t and r. With $\dot\theta = \dot\phi = 0$ they reduce to

$$\ddot t + \frac{r_s}{r(r - r_s)}\dot r \dot t = 0 \qquad \ddot r - \frac{1}{2}\frac{r_s}{r(r - r_s)}\dot r^2 + r_s\frac{r - r_s}{2r^3}\dot t^2 = 0\ ,$$

which is a special case $J = 0$ of what we had in lecture. We thus have

$$\left(1 - \frac{r_s}{r}\right)\dot t = H \qquad \frac{r}{r - r_s}(H^2 - \dot r^2) = K^2\ .$$

For $K = 1$ we have $p = \tau$ and thus

$$\frac{dr}{d\tau} = -\sqrt{H^2 - \left(1 - \frac{r_s}{r}\right)}$$

and

$$\frac{dr}{dt} = -\frac{1 - \frac{r_s}{r}}{H}\sqrt{H^2 - \left(1 - \frac{r_s}{r}\right)}\ .$$

(b) Demanding

$$\left.\frac{dr}{d\tau}\right|_{r\to\infty} = 0$$

implies $H = 1$, i.e.

$$\frac{dr}{d\tau} = -\sqrt{\frac{r_s}{r}} \qquad \frac{dr}{dt} = -\left(1 - \frac{r_s}{r}\right)\sqrt{\frac{r_s}{r}} .$$

We can now integrate

$$\tau = \int_{2r_s}^{r_s} \frac{d\tau}{dr} dr = -\frac{1}{\sqrt{r_s}} \int_{4r_s}^{r_s} \sqrt{r} \, dr = \frac{1}{\sqrt{r_s}} \frac{2}{3} r^{3/2}\big|_{r_s}^{4r_s} = \frac{14}{3} r_s ,$$

which is the eigentime it takes the infalling observer to reach r_s from $2r_s$. The same integral for the coordinate time results in

$$t = \int_{4r_s}^{r_s} \frac{dt}{dr} dr = -\frac{1}{\sqrt{r_s}} \int_{4r_s}^{r_s} \frac{r^{3/2}}{r - r_s} dr = \frac{1}{\sqrt{r_s}} \int_{r_s}^{4r_s} \frac{r^{3/2}}{r - r_s} dr ,$$

which diverges as $r \to r_s$. Since the coordinate time is the time of a distant observer, she will never see an infalling object reach the horizon.

2. (a) The first part of the exercise can be conveniently checked using one of the symbolic algebra routines in Appendix F. From this solution we need two Christoffel symbols for the later parts of the problem: $\Gamma^r_{tt} = f'(r)f(r)/2$ and $\Gamma^r_{\varphi\varphi} = -rf(r)\sin^2\theta$.

   (b) To find the position of a horizon, we locate a coordinate singularity as in the case of the Schwarzschild metric. From the metric we see that there is a coordinate singularity at $f(r) = 0$, so

$$1 - \frac{2MG}{r} - \frac{1}{3}\Lambda r^2 = 0 .$$

This is a cubic equation, which we will not solve exactly. Instead, we write it first as

$$1 - \frac{2MG}{r} = \frac{1}{3}\Lambda r^2 .$$

Now assume that the right hand side is a small correction. Then we can write the solution as

$$r_1 = 2MG + \varepsilon$$

and, plugging it in, we obtain that to first order in ε

$$1 - \frac{2MG}{2MG + \varepsilon} \approx \frac{\varepsilon}{2MG} \approx \frac{1}{3}\Lambda(2MG)^2 ,$$

which is small since we assumed $MG\sqrt{\Lambda} \ll 1$. Computing r_1 we obtain

$$r_1 = 2MG + \varepsilon \approx 2MG \left(1 + \frac{4M^2G^2}{3}\Lambda\right) .$$

Conversely, we may write $f(r) = 0$ as

$$1 - \frac{1}{3}\Lambda r^2 = \frac{2MG}{r}$$

and again treat the right hand side as a small correction. Now we can write the solution as

$$r_2^2 = \frac{3}{\Lambda} + \varepsilon \qquad r_2 \approx \sqrt{\frac{3}{\Lambda}}\left(1 + \frac{1}{2}\varepsilon\frac{\Lambda}{3}\right) .$$

With this ansatz we find

$$-\varepsilon\frac{\Lambda}{3} \approx -2MG\sqrt{\frac{\Lambda}{3}} ,$$

which again is small since we assumed $MG\sqrt{\Lambda} \ll 1$ and thus

$$r_2 \approx \sqrt{\frac{3}{\Lambda}}\left(1 - MG\sqrt{\frac{\Lambda}{3}}\right) .$$

(c) The geodesic equation is

$$\frac{du^\mu}{d\tau} = u^\nu\nabla_\nu u^\mu = 0 .$$

Since we assume the velocity to be constant in our coordinates, this simplifies to

$$u^\nu\nabla_\nu u^\mu = u^\nu\Gamma^\mu_{\nu\alpha}u^\alpha = 0 .$$

We choose coordinates such that the orbit is in the $\theta = \pi/2$ plane. The velocity vector written explicitly in spherical coordinates (t, r, θ, φ), is then

$$u^\mu = \begin{pmatrix} \alpha \\ 0 \\ 0 \\ \omega \end{pmatrix} ,$$

and the corresponding covariant vector is

$$u_\mu = \begin{pmatrix} f(r)\alpha \\ 0 \\ 0 \\ -r^2 \sin^2 \theta \omega \end{pmatrix} ,$$

Because we parameterised with the eigentime τ and $d\tau^2 = dx^\mu dx_\mu$, the velocity vector $u^\mu = dx^\mu/d\tau$ has unit norm. Thus

$$u^\mu u_\mu = f(r)\alpha^2 - r^2\omega^2 = 1$$

It turns out that the only nontrivial component in the geodesic equation is the r-component. The relevant Christoffel symbols for $\theta = \pi/2$ that do not vanish are

$$\Gamma^r_{\varphi\varphi} = -rf(r) \qquad \Gamma^r_{tt} = \frac{1}{2}f(r)f'(r)$$

and with those the geodesic equation reduces to the condition

$$\frac{\alpha^2}{2}f(r)f'(r) - \omega^2 rf(r) = 0 ,$$

plugging in the unit norm condition $f(r)\alpha^2 = 1 + r^2\omega^2$ we finally arrive at the solution

$$\omega^2 = \frac{1}{r\left(2\frac{f(r)}{f'(r)} - r\right)} .$$

The right hand side of this equation has to be positive, so we require

$$\frac{2f(r)}{f'(r)} > r$$

Let us first assume that $f'(r) > 0$. Plugging in $f(r) = 1 - 2GM/r - \Lambda r^2/3$, this condition translates into

$$r < \left(\frac{3MG}{\Lambda}\right)^{1/3} .$$

In that case, we have $2f(r) > f'(r)$ which implies

$$r > 3MG$$

So stable orbits can exist for radial coordinates

$$3MG < r < \left(\frac{3MG}{\Lambda}\right)^{1/3} .$$

Now let us suppose that $f'(r) < 0$. This would reverse both inequalities and we would obtain the condition

$$3MG > r > \left(\frac{3MG}{\Lambda}\right)^{1/3} ,$$

which obviously implies

$$3MG > \left(\frac{3MG}{\Lambda}\right)^{1/3} .$$

Rearranging this inequality we find

$$(3MG)^{2/3}\Lambda^{1/3} > 1$$

which contradicts our assumption $MG\sqrt{\Lambda} \ll 1$. So eventually stable orbits can be found in the region

$$3MG < r < \left(\frac{3MG}{\Lambda}\right)^{1/3} .$$

only.

3. The life time of a black hole due to Hawking radiation is approximately

$$t = \frac{5120\pi G^2}{\hbar} M_0^3 .$$

In SI units we have

$$\frac{5120\pi G^2}{\hbar} \simeq 6.8 \times 10^{17} \frac{m^4}{kg^3 \, s^3}$$

and since we used $c = 1$, we have

$$1m \simeq \frac{1}{3}10^{-8} \, s ,$$

which results in

$$\frac{5120\pi G^2}{\hbar} \simeq 8.4 \times 10^{-17} \frac{s}{kg^3} .$$

The mass of the Sun is about $M_\odot \simeq 2 \times 10^{30}$ kg, so we finally find

$$\frac{5120\pi G^2}{\hbar} \simeq 6.7 \times 10^{74} \frac{s}{M_\odot^3} = 2.1 \times 10^{67} \frac{yr}{M_\odot^3} \ ,$$

i.e. a black hole of one solar mass has a life time of $\sim 2.1 \times 10^{67}$ years, which is 57 orders of magnitude larger than the current age of our universe! Consequently, Sgr A* will decay in about $\sim 10^{87}$ and the central black hole of M87 in $\sim 10^{97}$ years. If on the other hand we demand the decay time to be the age of our universe $T \sim 1.37 \times 10^{10}$ yr, we find a mass of

$$\left(\frac{M_0}{M_\odot}\right)^3 \simeq \frac{1.37 \times 10^{10}}{2.1 \times 10^{67}} \simeq 6.4 \times 10^{-58}$$

or

$$M_0 \simeq 8.6 \times 10^{-20} M_\odot \simeq 1.7 \times 10^{11} \ \text{kg} \ .$$

This means that a black hole of this original mass would currently decay, provided it has not obtained any new energy since its creation.

E.6 Chapter 7

1. We use (7.2), i.e.

$$D = v/H_0 \ .$$

With $v = 1000 \, \text{km s}^{-1}$ we obtain $D \approx 50 \cdot 10^6$ lt yr. For comparison—the distance to Andromeda is $2.5 \cdot 10^6$ lt yr.

2. The velocity of the second observer due to the expansion is $\delta u = \frac{\dot{a}}{a}\delta l$ (cf. (7.2) and (7.4)). The particle arrives at the second observer at time δt, i.e.

$$\delta u = \frac{\dot{a}}{a} v \, \delta t = v \frac{\delta a}{a} \ .$$

According to the velocity addition theorem the second observer measures the particle velocity v' given by

$$v' = \frac{v - \delta u}{1 - v\delta u} = v \underbrace{-(1 - v^2)\delta u}_{=\delta v} + \mathcal{O}(\delta u^2) \ .$$

Combination of the two equations yields

$$\delta v = -v(1 - v^2)\frac{\delta a}{a} \ .$$

Integrating both sides we find

$$p \propto \frac{1}{a} \ .$$

E.7 Chapter 8

1. (a) Simple geometry yields $\delta l = R \sin(r/R)\delta\phi$.
   (b) The circumference is $S = 2\pi R \sin(r/R)$. Expansion of S yields $S \approx 2\pi r (1 - \frac{1}{6}(r/R)^2)$ and thus $\frac{1}{6}(r/R)^2) \approx 4 \cdot 10^{-9}$. This means that the precision of his measurement of the circumference must be about 10^{-2} mm.
   (c) Produce a good sketch and use that the 3D distance x between the end-points of the object can be expressed via $x^2 = R^2 + R^2 - 2R^2 \cos\phi$ or $x^2 = 2R^2 \sin^2\theta \, (1 - \cos\varphi)$.
2. (a) The coordinates are $x = \cos\varphi \sin\theta \sin\alpha$, $\quad y = \sin\varphi \sin\theta \sin\alpha$, $z = \cos\theta \sin\alpha$, and $w = \cos\alpha$.
   (b) The coordinates are $x = \cos\varphi \sinh\alpha$, $y = \sin\varphi \sinh\alpha$, and $z = \cosh\alpha$. Note that $x^2 + y^2 - z^2 = -1$. This is a hyperboloid of two sheets, which is shown in the figure below—including a unit sphere between the two sheets. We have added the sphere because it corresponds to the case of Ω_2. The sketch on the right is a hyperboloid of one sheet—also added here for comparison. The hyperboloid of one sheet has the equation $x^2 + y^2 - z^2 = +1$.

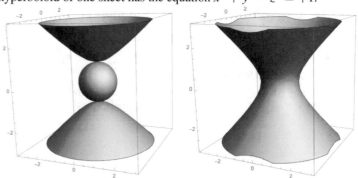

   (c) The expression is the same for a sphere of radius $a \sin\alpha$ in three-dimensional flat space. Thus, the total surface area is

$$A = 4\pi a^2 \sin^2\alpha \ .$$

   Note that as the radius α grows, the area initially grows, reaching a maximum value for $\alpha = \pi/2$, and then decreases to zero at $\alpha = \pi$. Because the physical width of an infinitesimal shell is $a \, d\alpha$, the volume element between to spheres with radii α and $\alpha + d\alpha$ is given by

$$dV = A \, a \, d\alpha \, .$$

Integration yields

$$V = \int_0^{\alpha_o} A \, a \, d\alpha = 2\pi a^3 \left(\alpha_o - \frac{1}{2} \sin(2\alpha_o) \right) \, .$$

In the limit $\alpha \to 0$ this reduces to the flat-space form, i.e.

$$V = 4\pi (a \, \alpha_o)^3 / 3 + \cdots$$

and in the limit $\alpha = \pi$ we find

$$V = 2\pi^2 a^3 \, .$$

(d) An analogous calculation yields

$$A = 4\pi a^2 \sinh^2 \alpha$$

and

$$V = 2\pi a^3 \left(-\alpha_o + \frac{1}{2} \sinh(2\alpha_o) \right) \, .$$

In the limit $\alpha \to 0$ this reduces again to the above flat-space form. V of course grows without bound when α grows.

3. The solution to this problem can be found in Appendix F. But you should show your own work!

4. The relation follows via combination of (8.34), i.e. $\dot\rho a^3 + 3(\rho + P)a^2 \dot a = 0$, with the first Friedmann equation, i.e. (8.26).

5. (a) From the first Friedmann equation we have

$$\dot a = \sqrt{\frac{8\pi G}{3} \rho a^2 - K}$$

or after integration

$$a(t) = \int_0^t dt' \sqrt{\frac{8\pi G}{3} \rho a^2 - K}$$

If $K = -1$ (open) or $K = 0$ (flat) the argument of the square root is always greater or equal to zero, which means that $a(t)$ increases with increasing t. The case $K = +1$ (closed) requires more care. If we assume $\rho \propto a^{-q}$ with $q > 0$, i.e. ρ decreases as a increases, we have $\rho a \propto a^{1-q}$. If $q < 1$ then the square root becomes complex when $a \to 0$ for $K > 0$. Thus we expect $q > 1$. Note that K is a constant and therefore is unimportant in the limit $a \to 0$ for $q > 1$. As a increases, the argument under the square root approaches zero. Thus the growth must come to a halt at some a_{max}. The condition $\rho + 3P > 0$ implies $\ddot{a} < 0$ according to the second Friedmann equation, which means that $\dot{a}$ is monotonously decreasing. In the case at hand $\dot{a}$ in fact is negative for times greater than the time when $a = a_{max}$. This is the collapse. Note that a monotonous decrease of $\dot{a}$ does not require $\dot{a} < 0$. If $K = -1$ then $\dot{a}$ approaches are limiting positive value, whereas for $K = 0$ it approaches zero.

(b) How is this different from part (a)? We find that $-\rho \le P \le -\rho/3$ implies $\ddot{a} \ge 0$ in this regime, whereas in (a) we had $\ddot{a} < 0$. It means that $a(t)$ undergoes a steady growth and no turn-around is possible, because this requires $\dot{a}$ to decrease and eventually become negative and thus $\ddot{a} < 0$. In other word—a closed universe is not possible in this regime.

(c) Here we find $\dot{\rho} > 0$ (note: $\dot{\rho} = -3H(\rho + P)$) for $H > 0$, which makes no physical sense.

6. We chose the second Friedmann equation as our starting point. Static means that $\ddot{a} = 0$ and thus $\rho + 3P = 0$. Because cosmological constant is synonymous with dark or vacuum energy may write $\rho = \rho_v + \rho_m$, where $P = -\rho_v$. Thus

$$0 = \rho_v + \rho_m - 3\rho_v = -2\rho_v + \frac{\text{const}}{a_o{}^3}$$

or

$$a_o = \left(\frac{\text{const}}{2\rho_v}\right)^{1/3}.$$

Now we add a small time-dependent perturbation, i.e. $a(t) = a_o + \delta a(t)$. Inserting this into the second Friedmann equation and keeping linear terms only yields

$$\delta\ddot{a}(t) = \frac{4\pi G\rho_m}{a_o}\delta a(t).$$

The general solution is $\delta a(t) = c_1 \exp[-\lambda t] + c_2 \exp[\lambda t]$ ($\lambda > 0$). If we require $\delta a(t)$ to be infinitesimally small at $t = 0$, we find $c_1 = -c_2$ and therefore $\delta a(t) \propto \sinh[\lambda t]$, which grows without bound as t increases.

7. We start from Newton's equation $-\vec{\nabla}\varphi = \ddot{\vec{r}}$, where we set $\vec{r}(t) = a(t)\vec{x}$. Application of $\vec{\nabla}$ to both sides of the equation and using $\vec{\nabla}\ddot{\vec{r}} = (\ddot{a}(t)/a(t))\vec{\nabla}_x \cdot \vec{x} = 3\ddot{a}(t)/a(t)$ yields the desired result.

E.8 Chapter 9

1. (a) The surface of the sphere is $A = 4.84\,\mathrm{m}^2$. Use $m_w C \mathrm{d}T/\mathrm{d}t = -\sigma_{SB}T^4 A$, where $m_W = 1000\,\mathrm{kg}$ is the mass of the water. Integration of this equation yields

$$T(t) = \left(T_i^{-3} + \frac{3\sigma_{SB}A}{m_w C}t\right)^{-1/3}.$$

Thus $T(10\,\mathrm{h}) \approx 281\,\mathrm{K}$.

(b) Here we have $\mathrm{d}T/\mathrm{d}t = \sigma_{SB}AT_{CMB}^4/(m_w C)$. With $T_{CMB} \approx 2.7\,\mathrm{K}$ we therefore obtain $\Delta T(10\,\mathrm{h}) \approx 1.3 \cdot 10^{-7}\,\mathrm{K}$.

2. According to (9.42) we have $\rho_{r,0}/\rho_{c,0} \approx 9 \cdot 10^{-5}$ and according to (9.48) $\rho_{b,0}/\rho_{c,0} \approx 0.05$. Using $\rho_{r,0}/\rho_{b,0} = (2.725\,\mathrm{K}/T)^4$ we obtain $T \approx 13.2\,\mathrm{K}$. This is considerably less than the surface temperature of a star.

3. Considering the office as a black body, we can describe the radiation pressure via $P_r = \frac{1}{3}\sigma T^4 \approx 1.9 \cdot 10^{-6}\,\mathrm{Pa}$. Therefore $P \approx P_m$ or $P_m/P_r \approx 5 \cdot 10^{10}$, i.e. $\eta_{\text{office}} \approx 10^{10}$!

4. We have $\mathrm{d}E = -P\mathrm{d}V$ and, expressing E as a function of T and V, we also have

$$\mathrm{d}E = \left.\frac{\partial E}{\partial V}\right|_T \mathrm{d}V + C_V\mathrm{d}T \overset{(9.9)}{=} \left(T\left.\frac{\partial P}{\partial T}\right|_V - P\right)\mathrm{d}V + C_V\mathrm{d}T.$$

The quantity C_V is the isochoric heat capacity $C_V = \partial E/\partial T|_V$. Thus

$$T\left.\frac{\partial P}{\partial T}\right|_V \mathrm{d}V + C_V\mathrm{d}T = 0.$$

Using $E = mN + \frac{3}{2}NT + \sigma VT^4$ together with $P_m = NT/V$ and $P_r = \frac{1}{3}\sigma T^4$ we obtain

$$\left(NT + \frac{4}{3}\sigma VT^4\right)\frac{\mathrm{d}V}{V} + \left(\frac{3}{2}NT + 4\sigma VT^4\right)\frac{\mathrm{d}T}{T} = 0.$$

The desired result now follows via $\mathrm{d}V/V = 3\mathrm{d}a/a$.

5. (a) Let's study an example. In the following a state is indicated via a dash $(-)$ and a Boson via a circle $(\circ)$. Suppose we have four degenerate and distinguishable states, i.e. $-\,-\,-\,-$, all possessing the energy ϵ_i, and three Bosons, i.e. $\circ\,\circ\,\circ$. A particular arrangement is $|-\circ\circ|-\circ|-|-$, which means that the first state is occupied by two Bosons, the second by one and the remaining two are unoccupied. This shows that all possible arrangements of g_i states and N_i Bosons are given by $(g_i - 1 + N_i)! \approx (g_i + N_i)!$ permutations. The -1 accounts for the one state which must come first. Because particles are indistinguishable, we divide by $N_i!$. We also must divide by $(g_i - 1)! \approx g_i!$, because the above permutations do include the states, which we do not want

to permute. Thus

$$B_i = \frac{(g_i + N_i)!}{g_i! N_i!} .$$

Using $S_B = \sum_i \ln B_i$ in conjunction with Stirling's approximation we have

$$S_B = \sum_i [(g_i + N_i) \ln(g_i + N_i) - g_i \ln g_i - N_i \ln N_i]$$

and with $n_i = N_i/g_i$

$$S_B = \sum_i g_i [(n_i + 1) \ln(n_i + 1) - n_i \ln n_i]$$

Variation of S_B amounts to finding the solution of

$$0 = \frac{d}{dn_k} \sum_i g_i [(n_i + 1) \ln(n_i + 1) - n_i \ln n_i + \lambda_1 \epsilon_i n_i + \lambda_2 n_i] ,$$

i.e.

$$n_k^{(B)} = (\exp[-\lambda_1 \epsilon_k - \lambda_2] - 1)^{-1} .$$

(b) Let's again look at an example. In the following a state is indicated as before via a dash (−) and a Fermion via a circle (∘). Suppose we have four degenerate and distinguishable states, i.e. − − −−, all possessing the energy ϵ_i, and three Fermions, i.e. ∘ ∘ ∘. A particular arrangement is | − | − ∘| − ∘| − ∘, which means that the first state is unoccupied and the other states are occupied by one particle. In the Fermion case we build all possible arrangements by inserting one circle between two neighboring dashes. This shows that all possible arrangements of g_i states and N_i Fermions are given by $(g_i − 1)(g_i − 2)...(g_i − (N_i − 1)) = g_i!/(g_i − N_i)!$. Because the particles are indistinguishable, we obtain

$$F_i = \frac{g_i!}{(g_i − N_i)! N_i!} .$$

Using $S_F = \sum_i \ln F_i$ in conjunction with Stirling's approximation we have

$$S_F = \sum_i [g_i \ln g_i - N_i \ln N_i - (g_i - N_i) \ln(g_i - N_i)]$$

and with $n_i = N_i/g_i$

$$S_F = \sum_i g_i \left[-(1 - n_i) \ln(1 - n_i) - n_i \ln n_i \right] .$$

The equilibrium occupation number follows via $0 = \frac{d}{dn_k}(S_F + \sum_i g_i [\lambda_1 \epsilon_i n_i + \lambda_2 n_i])$, i.e.

$$n_k^{(F)} = (\exp[-\lambda_1 \epsilon_k - \lambda_2] + 1)^{-1} .$$

(c) The summation over the states is rewritten via

$$\sum_i = \frac{V}{(2\pi)^3} \int d^3k = \frac{V}{2\pi^2 c^3} \int_0^\infty d\omega\, \omega^2 .$$

For the energy density we find

$$\frac{E}{V} = \frac{1}{V} \sum_i g_i \epsilon_i n_i = \frac{g\hbar}{2\pi^2 c^3} \int_0^\infty d\omega \frac{\omega^3}{e^{\beta\hbar\omega} \pm 1} ,$$

assuming that all states possess identical degeneracy g, or with the substitution $x = \beta\hbar\omega$

$$\frac{E}{V} = \frac{g}{2\pi^2 c^3 \hbar^3 \beta^4} \int_0^\infty dx \frac{x^3}{e^x \pm 1} .$$

Note that $\pm$ stands for Bosons ($-$) and Fermions ($+$).

E.9 Chapter 10

1. (a) In the case of model (i) numerical integration of the integral in (10.7) yields 0.964. For model (ii) the corresponding value is 1.003. The respective ages of the universe are $13.5 \cdot 10^9$ y and $14.0 \cdot 10^9$ y.

   (b) Again we want to use (10.7) and therefore we must know the value of z, which follows via

$$\frac{\Omega_m / \Omega_r}{1} = \frac{\rho_m(t_o)/\rho_r(t_o)}{\rho_m(t_{rm})/\rho_r(t_{rm})} = \frac{a(t_o)}{a(t_{rm})} = 1 + z .$$

   For model (i) we find $z = 3332$ whereas for model (ii) $z = 2888$. The respective cross-over times are $t_{rm} \approx 5.0 \cdot 10^4$ y after the Big Bang and $t_{rm} \approx 6.9 \cdot 10^4$ y after the Big Bang.

   (c) The calculation is analogous to the one in part (b). Here the value of z follows via

$$\frac{\Omega_v/\Omega_m}{1} = \frac{\rho_v(t_o)/\rho_m(t_o)}{\rho_v(t_{mv})/\rho_m(t_{mv})} = \frac{a(t_o)^3}{a(t_{mv})^3} = (1+z)^3 .$$

For model (i) we find $z = 0.326$ whereas for model (ii) $z = 0.417$. The respective cross-over times are $t_{mv} \approx 9.5 \cdot 10^9$ y after the Big Bang (which means about $4 \cdot 10^9$ y ago) and $t_{mv} \approx 9.6 \cdot 10^9$ y after the Big Bang (which means about $4.4 \cdot 10^9$ y ago).

2. We insert $\rho_m(t) \approx 6\rho_b(t)$ into

$$\frac{\rho_m(t_o)/\rho_\gamma(t_o)}{\rho_m(t)/\rho_\gamma(t)} = 1 + z .$$

(cf. the previous problem). Thus we find

$$\rho_b(t)/\rho_\gamma(t) \approx \frac{\Omega_m/\Omega_\gamma}{6(z(t)+1)} .$$

Setting $z(t_{ls}) = 1040$ and using $\Omega_m = 0.3$, $\Omega_\gamma = 9 \cdot 10^{-5} \cdot 2/(2+1.36) = 5.4 \cdot 10^{-5}$ yields $\rho_b(t_{ls})/\rho_\gamma(t_{ls}) \approx 0.9$.

3. (a) We have

$$l(t_{\text{today}}, 0) = \int_0^{\text{today}} \frac{dt}{a(t)} .$$

Using (10.13) with $\Omega_m = 1$ and all other Ω equal to zero this becomes

$$l(t_{\text{today}}, 0) = \frac{1}{H_o a_o} \int_0^1 \frac{dx}{\sqrt{x}} = \frac{2}{H_o a_o} .$$

or

$$D_{h,p}(0) = \frac{2}{H_o} .$$

(b) A numerical integration based on (10.12) yields

$$D_{h,p}(0) \approx \frac{3.24}{H_o} \approx 45.4 \cdot 10^9 \text{ lt yr} .$$

4. Use (10.7) and calculate numerically $\Delta t = t(z = \infty) - t(z = 2.5)$. You should get $\Delta t_{\Omega_m=1} \approx 1.43 \cdot 10^9$ y and $\Delta t_{\Omega_m=0.3, \Omega_v=0.7} \approx 2.58 \cdot 10^9$ y. Thus, if the age of the galaxy is older than $1.43 \cdot 10^9$ y, then the hypothesis is ruled out.

5. We find the inflection point in the lower panel of (10.12), via

$$0 = \frac{d^2}{dx^2} t(z) H_o = \frac{d}{dx} \frac{1}{\sqrt{\Omega_v x^2 + \Omega_m x^{-1}}} = \frac{2\Omega_v x - \Omega_m x^{-2}}{2(\Omega_v x^2 + \Omega_m x^{-1})^{3/2}} .$$

Thus

$$x_i = \left(\frac{\Omega_m}{2\Omega_v}\right)^{1/3} \approx 0.56 ,$$

i.e. $(1 + z_i)^{-1} \approx 0.599$ and therefore $z_i \approx 0.79$. Note: $t(z_i) \approx 7.1 \cdot 10^9$ y. Remark: For $\Omega_m = 0.3$ and $\Omega_v = 0.7$ one finds $z_i \approx 0.67$ and $t(z_i) \approx 7.3 \cdot 10^9$ y.

6. According to (10.21) and (10.22) we can write

$$D_{h,p}(t) = a(t) \int_0^t \frac{dt'}{a(t')} .$$

Next we make use of (10.6), i.e.

$$dt' = \frac{dx}{H_0 x \sqrt{\Omega_v + \cdots}} ,$$

where $x = a(t')/a_0 = 1/(1 + z(t'))$. Combination of the two formulas yields

$$D_{h,p}(t) = a(t) \int_0^{1/(1+z(t))} \frac{dx}{H_0 a_0 x^2 \sqrt{\Omega_v + \cdots}}$$

$$= \frac{1}{H_0(1 + z(t))} \int_0^{1/(1+z(t))} \frac{dx}{x^2 \sqrt{\Omega_v + \cdots}} . \qquad (E.1)$$

7. We have $T \propto a^{-1}$ and we also have $1 + z = a(t_o)/a(t_e)$. The combination immediately yields the desired relation, where $T(z) \propto a^{-1}(t_e)$ and $T_o \propto a^{-1}(t_o)$.

8. (a) We can compute $\Lambda = 8\pi G \rho_v \simeq 1.18 \times 10^{-52}$ m^{-2}. With $GM_\odot \simeq 1498$ m this results in $MG\sqrt{\Lambda} = 1.6 \times 10^{-23}$, so $MG\sqrt{\Lambda} \ll 1$ is an extremely good approximation.

   (b) From the solution of Problem 2 from Chap. 6 we know that the maximum allowed circular orbit is at a radial coordinate

$$r < \left(\frac{3MG}{\Lambda}\right)^{1/3} \simeq 3.35 \times 10^{18} \text{ m} .$$

This distance is over 100 pc or more than 300 lt yr, which is huge compared to the size of the entire solar system. This indicates that on solar system scales the cosmological constant is irrelevant.

   (c) For our local cluster of galaxies we find that the approximation $MG\sqrt{\Lambda} = 3.2 \times 10^{-11} \ll 1$ is still very good. For the largest stable orbit we obtain however

$$r < \left(\frac{3MG}{\Lambda}\right)^{1/3} \simeq 4.2 \times 10^{22} \text{ m} \simeq 1.36 \text{ Mpc}$$

which is comparable to the radius of the local group. This confirms that at galactic distances the cosmological constant definitely plays a role.

E.10 Chapter 11

1. Note that light rays travel at $45°$ angles with respect the two axes (as in Fig. 9.3). First we want to calculate the values of $\Delta_m \eta$ for the surface of last scattering and for a present observer in reference to the end of inflation at t_{ie}. Using $a(t) = a(t_{ie})(t/t_{ie})^{2/3}$ we find

$$\Delta_m \eta(t) = \frac{3t_{ie}^{2/3}}{a(t_{ie})}(t^{1/3} - t_{ie}^{2/3}) \approx \frac{3t_{ie}^{2/3}t^{1/3}}{a(t_{ie})} \approx \frac{3t_{ie}^{2/3}t^{1/3}}{e^{65}a(0)} .$$

During inflation, i.e. $a(t) \propto e^{Ht}$, we have

$$\Delta_i \eta(t) = \frac{\left(1 - e^{-Ht}\right)}{a(0)H} .$$

Here $a(0)$ is the scale factor at the beginning of inflation. With the values $Ht_{ie} = 65$ and $H^{-1} \approx 2.8 \cdot 10^{-38}$ s we obtain

$$\Delta_i \eta(t_{ie}) \approx \frac{1}{a(0)H} \approx 2.8 \cdot 10^{-38} \text{s} \frac{1}{a(0)} ,$$

Inserting for t in $\Delta_m \eta(t)$ the time of last scattering $t_{ls} \approx 4 \cdot 10^5$ y$\approx 1.3 \cdot 10^{13}$ s and the age of the universe $t_{today} \approx 4.4 \cdot 10^{17}$ s we obtain the three ratios

$$\frac{\Delta_m \eta(t_{today})}{\Delta_m \eta(t_{ls})} \approx 35 \quad \frac{\Delta_m \eta(t_{today})}{\Delta_i \eta(t_{ie})} \approx 7 \cdot 10^{-9} \quad \frac{\Delta_m \eta(t_{ls})}{\Delta_i \eta(t_{ie})} \approx 2 \cdot 10^{-10} .$$

In reference to Fig. 9.3 this means that the vertical separation of the 'now'-plane to the end-of-inflation-plane (or Big Bang-plane) is 35 times the vertical separation of the 'surfaces of last scattering'-plane to the end-of-inflation-plane. Thus, after the end of inflation, Fig. 9.3 is a reasonable sketch. What is completely different though is the Big Bang-plane itself. If its underside is the beginning of inflation and its upside is its end, then its thickness in the figure should be 10^8 times the vertical separation of the observer to the end of inflation. As a consequence, two points A and B anywhere on the surface of last scattering are easily brought into causal contact during inflation as is shown by the intersecting red lines in the sketch. This is another way how to look at the solution of the horizon problem

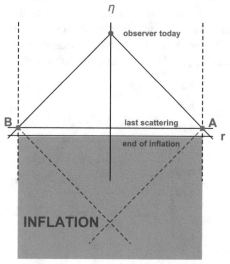

2. We begin by expressing the $A(\vec{x}_o, \vec{e}_i, \eta_i)$ in terms of their Fourier transforms, i.e.

$$C(\gamma) = \frac{1}{V} \int_V d^3 x_o \int \frac{d^3 k}{(2\pi)^3} \frac{d^3 k'}{(2\pi)^3} A_{\vec{k}} A_{\vec{k}'} e^{-i(\vec{k}+\vec{k}')\vec{x}_o} e^{-i(\vec{k}\vec{e}_1 \eta_1 + \vec{k}'\vec{e}_2 \eta_2)} .$$

Next we make use of the Fourier representation of the δ-function, i.e.

$$C(\gamma) = \frac{1}{V} \int \frac{d^3 k}{(2\pi)^3} d^3 k' \langle A_{\vec{k}} A_{\vec{k}'} \rangle \delta(\vec{k} + \vec{k}') e^{-i(\vec{k}\vec{e}_1 \eta_1 + \vec{k}'\vec{e}_2 \eta_2)}$$

$$= \frac{1}{V} \int \frac{d^3 k}{(2\pi)^3} \langle A_{\vec{k}} A_{-\vec{k}} \rangle e^{-i\vec{k}(\vec{e}_1 \eta_1 - \vec{e}_2 \eta_2)} .$$

In the next step we use $\vec{k}(\vec{e}_1 \eta_1 - \vec{e}_2 \eta_2) = k|\vec{e}_1 \eta_1 - \vec{e}_2 \eta_2| \cos \vartheta$, where $(\vec{e}_1 \eta_1 - \vec{e}_2 \eta_2)$ defines the current z-axis. Using spherical coordinates we obtain

$$C(\gamma) = \frac{1}{V} \int_0^\infty \frac{dk k^2}{(2\pi)^2} \langle A_{\vec{k}} A_{-\vec{k}} \rangle \int_0^\pi d\vartheta \sin \vartheta e^{-ik|\vec{e}_1 \eta_1 - \vec{e}_2 \eta_2| \cos \vartheta}$$

$$= \frac{1}{V} \int_0^\infty \frac{dk k^2}{2\pi^2} \langle A_{\vec{k}} A_{-\vec{k}} \rangle \frac{\sin(k|\vec{e}_1 \eta_1 - \vec{e}_2 \eta_2|)}{k|\vec{e}_1 \eta_1 - \vec{e}_2 \eta_2|} .$$

3. (a) We evaluate $C(\gamma)$ via

$$C(\gamma) = \frac{B}{2\pi^2} \sum_{l=1}^\infty \frac{2l + 1}{2l(l + 1)} P_l(\cos \gamma) .$$

The result is shown for ∞ replaced by $l_{max} = 10$ as well as $l_{max} = 200$ in the figure (dashed lines). Note that the negative $C(\gamma)$-values do not possess a special physical meaning, because we are free to shift our $C(\gamma)$ vertically. Both results coincide near $\gamma = \pi$. With decreasing γ they follow a logarithmic behavior. The range over which the logarithmic behavior is observed depends on how large l_{max} is. The solid line is the result of part (b).

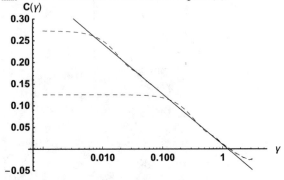

(b) We start via

$$P_\phi(r) = \frac{1}{(2\pi)^3} \int d^3k\, P_\phi(k) e^{-i\vec{k}\cdot\vec{r}} .$$

Following the same steps that lead to (11.47) we obtain

$$P_\phi(r) = \frac{V}{(2\pi)^2} B \int_{s_{min}}^{\infty} ds \frac{\sin s}{s^2} ,$$

where $s_{min} = k_{min}r = 2\pi r/L$. We note that the lower limit is problematic, because $s^{-2} \sin s \to s^{-1}$ in the limit of $s \to 0$. This means that the leading term will be $-\ln s_{min}$, i.e.

$$P_\phi(r) = \frac{V}{(2\pi)^2} B(-\ln(2\pi r/L) + \text{const}) .$$

The 'const' means that we let $s \to 0$ after having split off the divergent term. We can now compare this result to the above $C(\gamma)$. Here we assume that $\gamma \propto r$ in the limit of small r (and γ). The straight solid line in the figure is $P_\phi(r)/(VB)$, where we have adjusted the vertical position of $P_\phi(r)/(VB)$ using the constant. Note that we also absorb the term $-\ln(2\pi/L)$ into the constant.

E.11 Appendix D

1. The potential energy U of the dust cloud is proportional to the inverse of its radius R, i.e. $U \propto R^{-1}$. After virialization we have $U_f = -2K_f$, where K_f is the total kinetic energy of the dust cloud (which originally is at rest). Because the total energy $E = U_i = K_f + U_f = U_f/2$ is constant, we conclude $R_i^{-1} = 2R_f^{-1}$ or $R_f = R_i/2$.

2. (a) The general relativistic result for the precession angle per revolution is governed by the ratio of the gravitational potential energy GMm/R to the planet's rest mass mc^2, i.e.

$$\delta\phi \sim GM_\odot/c^2 R \sim 1.5 \text{ km}/58 \cdot 10^6 \text{ km} \sim 10^{-7} .$$

Here R is the average distance of the planet from the Sun. Dust distributed inside the orbit of the planet also leads to perihelion precession (cf. Problem 23 in [4]). In this case

$$\delta\phi \sim Gm\rho_{\text{dust}} R^3/(GmM) \sim M_{DM}/M_\odot \sim 10^{-19} .$$

Here ρ_{dust} is the dust's mass density, which we identify with Dark Matter, i.e. we use $\rho_{\text{dust}} \approx 0.5 \cdot 10^{-21} \text{ kg m}^{-3}$ for the Dark Matter density inside Mercury's orbit around the Sun and $M_\odot \sim 10^{30} \text{ kg}$. The result is that the 'perturbation' due to Dark Matter is insignificant.

(b) Note that the (plateau) Dark Matter density we have used in part (a) exceeds the average Dark Matter density in the universe by several orders of magnitude (cf. Figs. D.2 and D.3). Because today's Dark Energy and Dark Matter energy densities are of the same order of magnitude, we expect that the effect of Dark Energy is smaller by several orders of magnitude compared to the Dark Matter effect estimated in part (a). Thus, the 'perturbation' due to Dark Energy also is insignificant.

Appendix F
Algebraic Computer Codes

F.1 Mathematica Codes

The following is a compilation of algebraic code segments in Mathematica, which can be used to evaluate various quantities appearing in this text:

"Calculation of the Ricci tensor based on the Schwarzschild metric";

"metric tensor $g_{\nu\mu}$";

$g = \{\{\text{Exp}[2n[r]], 0, 0, 0\}, \{0, -\text{Exp}[2m[r]], 0, 0\}, \{0, 0, -r\char`^2, 0\},$

$\{0, 0, 0, -r\char`^2\text{Sin}[\theta]\char`^2\}\};$

Print $\left[\text{"}g_{\nu\mu} =\text{"}, \text{MatrixForm}[\%]\right]$

"metric tensor $g^{\nu\mu}$"; gInv = Inverse[g];

Print $[\text{"}g^{\nu\mu} =\text{"}, \text{MatrixForm}[\%]]$

"variables where t is time and speed of light is c=1";

$x = \{t, r, \theta, \phi\};$

"Christoffel symbols $\Gamma^{\sigma}_{\lambda\mu}$ (cf. (3.3.7) in Gravitation and

Cosmology)";

gam[σ_, λ_, μ_]:=

Sum[$(1/2)$gInv[[ν, σ]]

© Springer Nature Switzerland AG 2020
R. Hentschke and C. Hölbling, *A Short Course in General Relativity and Cosmology*, Undergraduate Lecture Notes in Physics,
https://doi.org/10.1007/978-3-030-46384-7

$(D[g[[\mu, \nu]], x[[\lambda]]] + D[g[[\lambda, \nu]], x[[\mu]]]-$

$D[g[[\mu, \lambda]], x[[\nu]]]), \{\nu, 1, 4\}];$

"calculation of the Ricci-tensor $R_{\nu\beta}$ (cf. B.73 on page 528

Cosmology - however with -sign)";

$RT = Table[Sum[D[gam[\alpha, v, b], x[[\alpha]]], \{\alpha, 1, 4\}]-$

$Sum[D[gam[\alpha, v, \alpha], x[[b]]], \{\alpha, 1, 4\}]-$

$Sum[gam[\alpha, v, d]gam[d, b, \alpha], \{\alpha, 1, 4\}, \{d, 1, 4\}]+$

$Sum[gam[\alpha, v, b]gam[d, d, \alpha], \{\alpha, 1, 4\}, \{d, 1, 4\}], \{v, 1, 4\},$

$\{b, 1, 4\}];$

$RTs = FullSimplify[RT];$

Print[" R_{00} =", RTs[[1, 1]]]

Print[" R_{11} =", RTs[[2, 2]]]

Print[" R_{22} =", RTs[[3, 3]]]

Print[" R_{33} =", RTs[[4, 4]]]

Print$\left[" R_{\nu\mu} =", 0, " otherwise"\right]$;

$$g_{\nu\mu} = \begin{pmatrix} e^{2n[r]} & 0 & 0 & 0 \\ 0 & -e^{2m[r]} & 0 & 0 \\ 0 & 0 & -r^2 & 0 \\ 0 & 0 & 0 & -r^2\mathrm{Sin}[\theta]^2 \end{pmatrix}$$

$$g^{\nu\mu} = \begin{pmatrix} e^{-2n[r]} & 0 & 0 & 0 \\ 0 & -e^{-2m[r]} & 0 & 0 \\ 0 & 0 & -\frac{1}{r^2} & 0 \\ 0 & 0 & 0 & -\frac{\mathrm{Csc}[\theta]^2}{r^2} \end{pmatrix}$$

$R_{00} = \frac{e^{-2m[r]+2n[r]}\left(n'[r](2-rm'[r]+rn'[r])+rn''[r]\right)}{r}$

$R_{11} = \frac{m'[r](2+rn'[r])-r(n'[r]^2+n''[r])}{r}$

$R_{22} = 1 + e^{-2m[r]}\left(-1 + rm'[r] - rn'[r]\right)$

$R_{33} = \mathrm{Sin}[\theta]^2\left(1 + e^{-2m[r]}\left(-1 + rm'[r] - rn'[r]\right)\right)$

$R_{\nu\mu} = 0$ otherwise

"Calculation of $\Gamma^\sigma_{\lambda\mu} dx^\lambda/d\tau\, dx^\mu/d\tau$ in the geodesic equations

of motion in the Schwarzschild metric";

"metric tensor $g_{\nu\mu}$";

$g = \{\{\text{Exp}[2n[r]], 0, 0, 0\}, \{0, -\text{Exp}[-2n[r]], 0, 0\},$

$\{0, 0, -r^\wedge 2, 0\}, \{0, 0, 0, -r^\wedge 2\text{Sin}[\theta]^\wedge 2\}\};$

"metric tensor $g^{\nu\mu}$";

gInv = Inverse[g];

"variables where t is time and speed of light is c=1";

$x = \{t, r, \theta, \phi\};$

"Christoffel symbols $\Gamma^\sigma_{\lambda\mu}$ (cf. (3.3.7) in Gravitation and

Cosmology)";

gam[$\sigma\_, \lambda\_, \mu\_$]:=

Sum[$(1/2)$gInv[[ν, σ]]

$(D[g[[\mu, \nu]], x[[\lambda]]] + D[g[[\lambda, \nu]], x[[\mu]]]-$

$D[g[[\mu, \lambda]], x[[\nu]]]), \{\nu, 1, 4\}];$

Do[Print["$\sigma = $", k, ": ",

Simplify[Sum[gam[k, i, j]Dt[$x[[i]], \tau$]Dt[$x[[j]], \tau$], $\{i, 1, 4\}$,

$\{j, 1, 4\}$]]/.θ->$\pi/2$], $\{k, 1, 4\}$]

$\sigma = 1$: 2Dt[r, τ]Dt[t, τ]$n'[r]$
$\sigma = 2$: $-e^{2n[r]}r$Dt[ϕ, τ]$^2 - \left(\text{Dt}[r, \tau]^2 - e^{4n[r]}\text{Dt}[t, \tau]^2\right)n'[r]$
$\sigma = 3$: 0
$\sigma = 4$: $\frac{2\text{Dt}[r,\tau]\text{Dt}[\phi,\tau]}{r}$

"Calculation of the Einstein tensor based on the FRW metric";

"metric tensor $g_{\nu\mu}$";

$g = \{\{1, 0, 0, 0\}, \{0, -a[t]^\wedge 2/(1 - kr^\wedge 2), 0, 0\},$

$\{0, 0, -a[t]^\wedge 2r^\wedge 2, 0\}, \{0, 0, 0, -a[t]^\wedge 2r^\wedge 2\text{Sin}[\theta]^\wedge 2\}\};$

```
Print [" g_{νμ} =", MatrixForm[%]]
```

```
"metric tensor g^{νμ} ";
```

```
gInv = Inverse[g];
```

```
Print [" g^{νμ} =", MatrixForm[%]]
```

"variables where t is time and speed of light is c=1";

$x = \{t, r, \theta, \phi\};$

"Christoffel symbols $\Gamma^{\sigma}_{\lambda\mu}$ (cf. (3.3.7) in Gravitation and

Cosmology)";

```
gam[σ_, λ_, μ_]:=
```

```
Sum[(1/2)gInv[[ν, σ]]
```

```
(D[g[[μ, ν]], x[[λ]]] + D[g[[λ, ν]], x[[μ]]]−
```

```
D[g[[μ, λ]], x[[ν]]]), {ν, 1, 4}];
```

"calculation of the Ricci-tensor $R_{\nu\beta}$ (cf. B.73 on page 528

Cosmology - however with -sign)";

```
RT = Table[Sum[D[gam[α, v, b], x[[α]]], {α, 1, 4}]−
```

```
Sum[D[gam[α, v, α], x[[b]]], {α, 1, 4}]−
```

```
Sum[gam[α, v, d]gam[d, b, α], {α, 1, 4}, {d, 1, 4}]+
```

```
Sum[gam[α, v, b]gam[d, d, α], {α, 1, 4}, {d, 1, 4}], {v, 1, 4},
```

```
{b, 1, 4}];
```

```
RTs = Simplify[RT];
```

```
Print [" R_{00} =", RTs[[1, 1]]] ;
```

```
Print [" R_{11} =", RTs[[2, 2]]] ;
```

```
Print [" R_{22} =", RTs[[3, 3]]] ;
```

```
Print [" R_{33} =", RTs[[4, 4]]] ;
```

```
Print [" R_{νμ} =", 0, " otherwise"] ;
```

"calculation of the Ricci-scalar R";

```
R = Sum[gInv[[i, j]]RT[[j, i]], {j, 1, 4}, {i, 1, 4}];
```

Simplify[R];

Print["R =", %]

"Einstein tensor $G_{\mu\nu}$";

G = **Simplify[Table[RT[[m, v]] − (1/2)g[[m, v]]R, {m, 1, 4},**

{v, 1, 4}]];

Print ["G_{00} =", G[[1, 1]]];

Print ["G_{11} =", G[[2, 2]]];

Print ["G_{22} =", G[[3, 3]]];

Print ["G_{33} =", G[[4, 4]]];

Print ["$G_{\nu\mu}$ =", 0, " otherwise"];

"Einstein tensor $G^{\mu\nu}$";

Gc = Table[Sum[gInv[[j, l]](gInv[[i, p]]G[[p, l]]), {p, 1, 4},

{l, 1, 4}], {i, 1, 4}, {j, 1, 4}];

Print ["G^{00} =", Gc[[1, 1]]]

Print ["G^{11} =", Gc[[2, 2]]]

Print ["G^{22} =", Gc[[3, 3]]]

Print ["G^{33} =", Gc[[4, 4]]]

Print ["$G^{\nu\mu}$ =", 0, " otherwise"];

$$g_{\nu\mu} = \begin{pmatrix} 1 & 0 & 0 & 0 \\ 0 & -\frac{a[t]^2}{1-kr^2} & 0 & 0 \\ 0 & 0 & -r^2a[t]^2 & 0 \\ 0 & 0 & 0 & -r^2a[t]^2\mathrm{Sin}[\theta]^2 \end{pmatrix}$$

$$g^{\nu\mu} = \begin{pmatrix} 1 & 0 & 0 & 0 \\ 0 & -\frac{1-kr^2}{a[t]^2} & 0 & 0 \\ 0 & 0 & -\frac{1}{r^2a[t]^2} & 0 \\ 0 & 0 & 0 & -\frac{\mathrm{Csc}[\theta]^2}{r^2a[t]^2} \end{pmatrix}$$

$$R_{00} = -\frac{3a''[t]}{a[t]}$$

$$R_{11} = \frac{2k + 2a'[t]^2 + a[t]a''[t]}{1 - kr^2}$$

$$R_{22} = r^2 \left(2k + 2a'[t]^2 + a[t]a''[t] \right)$$

$$R_{33} = r^2 \mathrm{Sin}[\theta]^2 \left(2k + 2a'[t]^2 + a[t]a''[t] \right)$$

$$R_{\nu\mu} = 0 \text{ otherwise}$$

$$R = - \frac{6\left(k + a'[t]^2 + a[t]a''[t]\right)}{a[t]^2}$$

$$G_{00} = \frac{3\left(k + a'[t]^2\right)}{a[t]^2}$$

$$G_{11} = \frac{k + a'[t]^2 + 2a[t]a''[t]}{-1 + kr^2}$$

$$G_{22} = - r^2 \left(k + a'[t]^2 + 2a[t]a''[t] \right)$$

$$G_{33} = - r^2 \mathrm{Sin}[\theta]^2 \left(k + a'[t]^2 + 2a[t]a''[t] \right)$$

$$G_{\nu\mu} = 0 \text{ otherwise}$$

$$G^{00} = \frac{3\left(k + a'[t]^2\right)}{a[t]^2}$$

$$G^{11} = \frac{\left(1 - kr^2\right)^2 \left(k + a'[t]^2 + 2a[t]a''[t]\right)}{\left(-1 + kr^2\right)a[t]^4}$$

$$G^{22} = - \frac{k + a'[t]^2 + 2a[t]a''[t]}{r^2 a[t]^4}$$

$$G^{33} = - \frac{\mathrm{Csc}[\theta]^2 \left(k + a'[t]^2 + 2a[t]a''[t]\right)}{r^2 a[t]^4}$$

$$G^{\nu\mu} = 0 \text{ otherwise}$$

"Numerical evaluation of (10.12) with

Ωo=0.7,ΩK=0,Ωm=0.3,Ωr=0 (black curve) and

Ωo=0.3,ΩK=0,Ωm=0.7,Ωr=0 (red curve)";

itime[Ωo_, ΩK_, Ωm_, Ωr_]:=

NIntegrate[1/(x^2Sqrt[Ωo + ΩK/x^2 + Ωm/x^3 + Ωr/x^4]),

{x, 1/(1 + z), 1}]

OK = 0.000001;

t1 =

Table[

{z, (1 + z)Sinh[Sqrt[OK]itime[0.7, OK, 0.3, 0.0]]/Sqrt[OK]},

{z, 0.01, 1, 0.01}];

t2 =

Table[

{z, (1 + z)Sinh[Sqrt[OK]itime[0.3, OK, 0.7, 0.0]]/Sqrt[OK]}, {z, 0.01, 1, 0.01}];

ListLogLogPlot[{t1, t2}, Joined → True,

AxesStyle → {{Black, Thickness[0.004]},

{Black, Thickness[0.004]}},

PlotStyle → {{Thickness[0.004], Black},

{Thickness[0.004], Red}}, AxesLabel → {z, " $H_0 d_L$ "}]

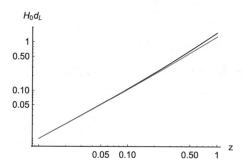

"Example shown in Fig. 10.1";

LogLogPlot[{$z + z^2, z + z^2/2$}, {z, 0.01, 1},

AxesStyle → {{Black, Thickness[0.004]},

{Black, Thickness[0.004]}},

PlotStyle → {{Thickness[0.004], Black},

{Thickness[0.004], Red}}, AxesLabel → {z, " $H_0 d_L$ "}]

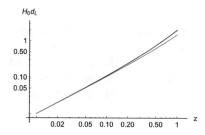

"Example shown in Fig. 10.4 (here A=a(t)/a(0))";

$A = 1$;

itime[$\Omega o\_, \Omega K\_, \Omega m\_, \Omega r\_$]:=

NIntegrate[$1/(x\,\text{Sqrt}[\Omega o + \Omega K/x^2 + \Omega m/x^3 + \Omega r/x^4])$, {x, 0, A}]

Clear[A];

data = Table[{itime[0.74, 0.0, 0.26, 0.00005], A},

{A, 0.01, 2, 0.001}];

ListPlot[{data, {{1, 1}}}, Joined → {True, True},

AxesStyle → {{Black, Thickness[0.003]},

{Black, Thickness[0.003]}},

PlotStyle → {{Thickness[0.005], Black},

{Thickness[0.03], Red}}, AxesLabel → {" H_0t", "a(t)/a(0)"}]

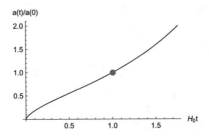

F.2   Maxima Codes

Here we provide some code to compute Christoffel symbols, the Ricci and Riemann curvature tensors and the Einstein tensor of an arbitrary metric using the free computer algebra system Maxima. The user needs to define a set of coordinates and the metric, which are defined in the first two statements. For the generic metric (5.1) that we used to derive the Schwarzschild solution the code is:

```
/* The coordinates */
coord:[t,r,theta,phi]$

/* Enter the metric */
metric:matrix([exp(2*n(r)),0,0,0],
              [0,-exp(2*m(r)),0,0],
              [0,0,-r^2,0],
              [0,0,0,-r^2*sin(theta)^2]);

/* Compute the inverse */
invmetric:invert(metric);

/* Define Christoffel symbols and print the nontrivial ones */
array(Gamma,4,4,4)$
for mu:1 thru 4 do for alpha:1 thru 4 do for beta:1 thru 4 do(
 Gamma[mu,alpha,beta]:factorsum(
  sum(
   invmetric[mu,gamma]*
   (diff(metric[alpha,gamma],coord[beta],1)
   +diff(metric[gamma,beta],coord[alpha],1)
   -diff(metric[alpha,beta],coord[gamma],1))/2,gamma,1,4)),
 if (Gamma[mu,alpha,beta]#0) and (alpha>=beta) then
 print(%Gamma[coord[alpha],coord[beta]]^coord[mu],"=",Gamma[mu,alpha,beta])
);

/* Define the Riemann tensor and print its nontrivial elements */
array(Riemann,4,4,4,4)$
for mu:1 thru 4 do for tau:1 thru 4 do for nu:1 thru 4 do for sigma:1 thru 4 do (
 Riemann[mu,tau,nu,sigma]:factorsum(
  diff(Gamma[mu,sigma,tau],coord[nu])
 -diff(Gamma[mu,nu,tau],coord[sigma])
 +sum(Gamma[mu,nu,rho]*Gamma[rho,sigma,tau]
    -Gamma[mu,sigma,rho]*Gamma[rho,nu,tau],rho,1,4)),
 if (Riemann[mu,tau,nu,sigma]#0) and (nu>sigma) then
 print(R[coord[tau],coord[nu],coord[sigma]]^coord[mu],"=",Riemann[mu,tau,nu,sigma])
);

/* Contract Riemann to Ricci tensor and print its nontrivial elements */
array(Ricci,4,4)$
for tau:1 thru 4 do for sigma:1 thru 4 do(
 Ricci[tau,sigma]:factorsum(sum(Riemann[mu,tau,mu,sigma],mu,1,4)),
 if (Ricci[tau,sigma]#0) and (tau>=sigma) then
 print(R[coord[tau],coord[sigma]],"=",Ricci[tau,sigma])
);

/* Compute Ricci scalar */
Riccis:factorsum(sum(sum(invmetric[mu,nu]*Ricci[mu,nu],mu,1,4),nu,1,4));

/* Compute the Einstein tensor and print its nontrivial elements*/
array(Einstein,4,4)$
for tau:1 thru 4 do for sigma:1 thru 4 do(
 Einstein[tau,sigma]:factorsum(Ricci[tau,sigma]-Riccis*metric[tau,sigma]/2),
 if (Einstein[tau,sigma]#0) and (tau>=sigma) then
 print(G[coord[tau],coord[sigma]],"=",Einstein[tau,sigma])
);
```

Running this code produces the following output:

$$\text{(metric)} \quad \begin{pmatrix} e^{2\,\mathrm{n}(r)} & 0 & 0 & 0 \\ 0 & -e^{2\,\mathrm{m}(r)} & 0 & 0 \\ 0 & 0 & -r^2 & 0 \\ 0 & 0 & 0 & -r^2 \sin(\theta)^2 \end{pmatrix}$$

$$\text{(invmetric)} \quad \begin{pmatrix} e^{-2\,\mathrm{n}(r)} & 0 & 0 & 0 \\ 0 & -e^{-2\,\mathrm{m}(r)} & 0 & 0 \\ 0 & 0 & -\frac{1}{r^2} & 0 \\ 0 & 0 & 0 & -\frac{1}{r^2 \sin(\theta)^2} \end{pmatrix}$$

$$\Gamma^{t}_{r,t} = \frac{\mathrm{d}}{\mathrm{d}r}\,\mathrm{n}(r)$$

$$\Gamma^{r}_{t,t} = e^{2\,\mathrm{n}(r)-2\,\mathrm{m}(r)}\left(\frac{\mathrm{d}}{\mathrm{d}r}\,\mathrm{n}(r)\right)$$

$$\Gamma^{r}_{r,r} = \frac{\mathrm{d}}{\mathrm{d}r}\,\mathrm{m}(r)$$

$$\Gamma^{r}_{\theta,\theta} = -r\,e^{-2\,\mathrm{m}(r)}$$

$$\Gamma^{r}_{\phi,\phi} = -r\,e^{-2\,\mathrm{m}(r)}\sin(\theta)^2$$

$$\Gamma^{\theta}_{\theta,r} = \frac{1}{r}$$

$$\Gamma^{\theta}_{\phi,\phi} = -\cos(\theta)\sin(\theta)$$

$$\Gamma^{\phi}_{\phi,r} = \frac{1}{r}$$

$$\Gamma^{\phi}_{\phi,\theta} = \frac{\cos(\theta)}{\sin(\theta)}$$

$$R^{t}_{r,r,t} = \frac{d^2}{dr^2}\,\mathrm{n}(r) + \left(\frac{\mathrm{d}}{\mathrm{d}r}\,\mathrm{n}(r)\right)\left(\frac{\mathrm{d}}{\mathrm{d}r}\,\mathrm{n}(r) - \frac{\mathrm{d}}{\mathrm{d}r}\,\mathrm{m}(r)\right)$$

$$R^{t}_{\theta,\theta,t} = r\,e^{-2\,\mathrm{m}(r)}\left(\frac{\mathrm{d}}{\mathrm{d}r}\,\mathrm{n}(r)\right)$$

$$R^{t}_{\phi,\phi,t} = r\,e^{-2\,\mathrm{m}(r)}\left(\frac{\mathrm{d}}{\mathrm{d}r}\,\mathrm{n}(r)\right)\sin(\theta)^2$$

$$R^{r}_{t,r,t} = e^{2\,\mathrm{n}(r)-2\,\mathrm{m}(r)}\left(\frac{d^2}{dr^2}\,\mathrm{n}(r) + \left(\frac{\mathrm{d}}{\mathrm{d}r}\,\mathrm{n}(r)\right)\left(\frac{\mathrm{d}}{\mathrm{d}r}\,\mathrm{n}(r) - \frac{\mathrm{d}}{\mathrm{d}r}\,\mathrm{m}(r)\right)\right)$$

$$R^{r}_{\theta,\theta,r} = -r\,e^{-2\,\mathrm{m}(r)}\left(\frac{\mathrm{d}}{\mathrm{d}r}\,\mathrm{m}(r)\right)$$

$$R^{r}_{\phi,\phi,r} = -r\,e^{-2\,\mathrm{m}(r)}\left(\frac{\mathrm{d}}{\mathrm{d}r}\,\mathrm{m}(r)\right)\sin(\theta)^2$$

$$R^{\theta}_{t,\theta,t} = \frac{e^{2\,\mathrm{n}(r)-2\,\mathrm{m}(r)}\left(\frac{\mathrm{d}}{\mathrm{d}r}\,\mathrm{n}(r)\right)}{r}$$

$$R^{\theta}_{r,\theta,r} = \frac{\frac{\mathrm{d}}{\mathrm{d}r}\,\mathrm{m}(r)}{r}$$

$$R^{\theta}_{\phi,\phi,\theta} = \left(e^{-\mathrm{m}(r)} - 1\right)\left(e^{-\mathrm{m}(r)} + 1\right)\sin(\theta)^2$$

$$R^{\phi}_{t,\phi,t} = \frac{e^{2\,\mathrm{n}(r)-2\,\mathrm{m}(r)}\left(\frac{\mathrm{d}}{\mathrm{d}r}\,\mathrm{n}(r)\right)}{r}$$

$$R^{\phi}_{r,\phi,r} = \frac{\frac{\mathrm{d}}{\mathrm{d}r}\,\mathrm{m}(r)}{r}$$

$$R^{\phi}_{\theta,\phi,\theta} = -\left(e^{-\mathrm{m}(r)} - 1\right)\left(e^{-\mathrm{m}(r)} + 1\right)$$

$$R_{t,t} = \frac{e^{2\,\mathrm{n}(r)-2\,\mathrm{m}(r)}\left(r\left(\frac{d^2}{dr}\,\mathrm{n}(r)\right) + r\left(\frac{\mathrm{d}}{\mathrm{d}r}\,\mathrm{n}(r)\right)^2 - r\left(\frac{\mathrm{d}}{\mathrm{d}r}\,\mathrm{m}(r)\right)\left(\frac{\mathrm{d}}{\mathrm{d}r}\,\mathrm{n}(r)\right) + 2\left(\frac{\mathrm{d}}{\mathrm{d}r}\,\mathrm{n}(r)\right)\right)}{r}$$

$$R_{r,r} = -\frac{r\left(\frac{d^2}{dr}\,\mathrm{n}(r)\right) + r\left(\frac{\mathrm{d}}{\mathrm{d}r}\,\mathrm{n}(r)\right)^2 - r\left(\frac{\mathrm{d}}{\mathrm{d}r}\,\mathrm{m}(r)\right)\left(\frac{\mathrm{d}}{\mathrm{d}r}\,\mathrm{n}(r)\right) - 2\left(\frac{\mathrm{d}}{\mathrm{d}r}\,\mathrm{m}(r)\right)}{r}$$

$$R_{\theta,\theta} = 1 - e^{-2\,\mathrm{m}(r)}\left(r\left(\frac{\mathrm{d}}{\mathrm{d}r}\,\mathrm{n}(r) - \frac{\mathrm{d}}{\mathrm{d}r}\,\mathrm{m}(r)\right) + 1\right)$$

$$R_{\phi,\phi} = -\left(e^{-2\,\mathrm{m}(r)}\left(r\left(\frac{\mathrm{d}}{\mathrm{d}r}\,\mathrm{n}(r) - \frac{\mathrm{d}}{\mathrm{d}r}\,\mathrm{m}(r)\right) + 1\right) - 1\right)\sin(\theta)^2$$

$$(\text{Riccis})\frac{2\left(e^{-2\,m(r)}\left(r\left(r\left(\frac{d^2}{dr^2}\,n(r)\right)+r\left(\frac{d}{dr}\,n(r)\right)^2-r\left(\frac{d}{dr}\,m(r)\right)\left(\frac{d}{dr}\,n(r)\right)+2\left(\frac{d}{dr}\,n(r)\right)-2\left(\frac{d}{dr}\,m(r)\right)\right)+1\right)-1\right)}{r^2}$$

$$G_{t,t}=-\frac{e^{2\,n(r)}\left(e^{-2\,m(r)}\left(-2r\left(\frac{d}{dr}\,m(r)\right)+1\right)-1\right)}{r^2}$$

$$G_{r,r}=\frac{e^{2\,m(r)}\left(e^{-2\,m(r)}\left(2r\left(\frac{d}{dr}\,n(r)\right)+1\right)-1\right)}{r^2}$$

$$G_{\theta,\theta}=r\,e^{-2\,m(r)}\left(r\left(\frac{d^2}{dr^2}\,n(r)\right)+r\left(\frac{d}{dr}\,n(r)\right)^2-r\left(\frac{d}{dr}\,m(r)\right)\left(\frac{d}{dr}\,n(r)\right)+\frac{d}{dr}\,n(r)-\frac{d}{dr}\,m(r)\right)$$

$$G_{\phi,\phi}=r\,e^{-2\,m(r)}\left(r\left(\frac{d^2}{dr^2}\,n(r)\right)+r\left(\frac{d}{dr}\,n(r)\right)^2-r\left(\frac{d}{dr}\,m(r)\right)\left(\frac{d}{dr}\,n(r)\right)+\frac{d}{dr}\,n(r)-\frac{d}{dr}\,m(r)\right)\sin(\theta)^2$$

For the Vaidya metric (6.37), the first two statements in the code have to be replaced by

```
/* The coordinates */
coord:[v,r,theta,phi]$

/* Enter the metric */
metric:matrix([1-2*G*m(v)/r,-1,0,0],
              [-1,0,0,0],
              [0,0,-r^2,0],
              [0,0,0,-r^2*sin(theta)^2]);
```

which results in the output:

$$(\text{metric})\begin{pmatrix}1-\frac{2G\,m(v)}{r} & -1 & 0 & 0\\ -1 & 0 & 0 & 0\\ 0 & 0 & -r^2 & 0\\ 0 & 0 & 0 & -r^2\sin(\theta)^2\end{pmatrix}$$

$$(\text{invmetric})\begin{pmatrix}0 & -1 & 0 & 0\\ -1 & \frac{2G\,m(v)}{r}-1 & 0 & 0\\ 0 & 0 & -\frac{1}{r^2} & 0\\ 0 & 0 & 0 & -\frac{1}{r^2\sin(\theta)^2}\end{pmatrix}$$

$$\Gamma^v_{v,v}=\frac{G\,m(v)}{r^2}$$

$$\Gamma^v_{\theta,\theta}=-r$$

$$\Gamma^v_{\phi,\phi}=-r\sin(\theta)^2$$

$$\Gamma^r_{v,v}=\frac{G\left(r^2\left(\frac{d}{dv}\,m(v)\right)-2G\,m(v)^2+r\,m(v)\right)}{r^3}$$

$$\Gamma^r_{r,v}=-\frac{G\,m(v)}{r^2}$$

$$\Gamma^r_{\theta,\theta}=2G\,m(v)-r$$

$$\Gamma^r_{\phi,\phi}=\sin(\theta)^2\left(2G\,m(v)-r\right)$$

$$\Gamma^\theta_{\theta,r}=\frac{1}{r}$$

$$\Gamma^\theta_{\phi,\phi}=-\cos(\theta)\sin(\theta)$$

$$\Gamma^\phi_{\phi,r}=\frac{1}{r}$$

$$\Gamma^\phi_{\phi,\theta}=\frac{\cos(\theta)}{\sin(\theta)}$$

$$R^{v}_{v,r,v} = -\frac{2G\,m(v)}{r^3}$$

$$R^{v}_{\theta,\theta,v} = \frac{G\,m(v)}{r}$$

$$R^{v}_{\phi,\phi,v} = \frac{G\sin(\theta)^2 m(v)}{r}$$

$$R^{r}_{v,r,v} = \frac{2G\,m(v)\,(2G\,m(v) - r)}{r^4}$$

$$R^{r}_{r,r,v} = \frac{2G\,m(v)}{r^3}$$

$$R^{r}_{\theta,\theta,v} = -G\left(\frac{\mathrm{d}}{\mathrm{d}v}m(v)\right)$$

$$R^{r}_{\theta,\theta,r} = \frac{G\,m(v)}{r}$$

$$R^{r}_{\phi,\phi,v} = -G\sin(\theta)^2\left(\frac{\mathrm{d}}{\mathrm{d}v}m(v)\right)$$

$$R^{r}_{\phi,\phi,r} = \frac{G\sin(\theta)^2 m(v)}{r}$$

$$R^{\theta}_{v,\theta,v} = \frac{G\left(r^2\left(\frac{\mathrm{d}}{\mathrm{d}v}m(v)\right) - 2G\,m(v)^2 + r\,m(v)\right)}{r^4}$$

$$R^{\theta}_{v,\theta,r} = -\frac{G\,m(v)}{r^3}$$

$$R^{\theta}_{r,\theta,v} = -\frac{G\,m(v)}{r^3}$$

$$R^{\theta}_{\phi,\phi,\theta} = -\frac{2G\sin(\theta)^2 m(v)}{r}$$

$$R^{\phi}_{v,\phi,v} = \frac{G\left(r^2\left(\frac{\mathrm{d}}{\mathrm{d}v}m(v)\right) - 2G\,m(v)^2 + r\,m(v)\right)}{r^4}$$

$$R^{\phi}_{v,\phi,r} = -\frac{G\,m(v)}{r^3}$$

$$R^{\phi}_{r,\phi,v} = -\frac{G\,m(v)}{r^3}$$

$$R^{\phi}_{\theta,\phi,\theta} = \frac{2G\,m(v)}{r}$$

$$R_{v,v} = \frac{2G\left(\frac{\mathrm{d}}{\mathrm{d}v}m(v)\right)}{r^2}$$

(Riccis)0

$$G_{v,v} = \frac{2G\left(\frac{\mathrm{d}}{\mathrm{d}v}m(v)\right)}{r^2}$$

References

1. A. Guth, Was cosmic inflation the 'Bang' of the Big Bang? Beam Line **27**, 14 (1997)
2. A. A. Michelson, The relative motion of the earth and the luminiferous ether. Am. J. Sci. **22**, 120 (1881)
3. A. A. Michelson, E.W. Morley, On the relative motion of the Earth and the luminiferous ether. Am. J. Sci. **34**, 333 (1887)
4. R. Hentschke, *Classical Mechanics* (Springer, Cham, 2017)
5. J.C. Haefele, R.E. Keating, Around-the-world atomic clocks: predicted relativistic time gains. Science **177**, 166 (1972)
6. R.V. Pound, G.A. Rebka Jr., Apparent weight of photons. Phys. Rev. Lett. **4**, 338 (1960)
7. R.V. Pound, G.A. Rebka Jr., Gravitational redshift in nuclear resonance. Phys. Rev. Lett. **3**, 439 (1959)
8. H.J. Hay, J.P. Schiffer, T.E. Cranshaw, P.A. Egelstaff, Measurement of the red shift in an accelerated system using the Mössbauer effect in Fe^{57}. Phys. Rev. Lett. **4**, 165 (1960)
9. J.D. Jackson, *Classical Electrodynamics* (Wiley, New York, 1975)
10. J. Wheeler, in *The Role of Gravitation in Physics, Report from the 1957 Chapel Hill Conference*, ed. by C.M. DeWitt, D. Rickles, Edition Open Access 2011 from the Max Planck Research Library for the History and Development of Knowledge
11. S. Weinberg, *Gravitation and Cosmology* (Wiley, New York, 1972)
12. L.D. Landau, E.M. Lifshitz, *The Classical Theory of Fields* (Butterworth-Heinemann, Oxford, 1975)
13. M. Laine, A. Vuorinen, Basics of Thermal Field Theory. Lecture Notes in Physics **1**, 925 (2016)
14. F. Hoyle, A new model for the expanding universe. Mon. Not. R. Astron. Soc. **108**, 372 (1948)
15. A.G. Riess, S. Casertano, W. Yuan, L.M. Macri, D. Scolnic, Large magellanic cloud cepheid standards provide a 1% foundation for the determination of the hubble constant and stronger evidence for physics beyond Λ CDM. Astrophys. J. **876**, 85 (2019)
16. W.L. Freedman et al., The Carnegie-Chicago hubble program. VIII. An independent determination of the hubble constant based on the tip of the red giant branch (2019), arXiv:1907.05922 [astro-ph.CO]
17. K.C. Wong et al., H0LiCOW XIII. A 2.4% measurement of H_0 from lensed quasars: 5.3σ tension between early and late-universe probes (2019), arXiv:1907.04869 [astro-ph.CO]
18. N. Aghanim et al., *[Planck Collaboration] Planck 2018 results* (VI, Cosmological parameters, 2018). arXiv:1807.06209 [astro-ph.CO]
19. N. Schöneberg, J. Lesgourgues, D.C. Hooper, The BAO+BBN take on the hubble tension (2019), arXiv:1907.11594 [astro-ph.CO]

© Springer Nature Switzerland AG 2020
R. Hentschke and C. Hölbling, *A Short Course in General Relativity and Cosmology*, Undergraduate Lecture Notes in Physics,
https://doi.org/10.1007/978-3-030-46384-7

20. J.C. Mather et al., A preliminary measurement of the cosmic microwave background (CMB) spectrum by the COsmic background explorer (COBE) Satellite. Astrophys. J. **354**, L37 (1990)
21. A.H. Guth, *The Inflationary Universe* (Random House, London, 1997)
22. R. Hentschke, *Thermodynamics* (Springer, Cham, 2017)
23. J. Overduin, H.-J. Fahr, Matter, spacetime and vacuum. Naturwissenschaften **88**, 491 (2001)
24. S. Weinberg, *The First Three Minutes* (Basic Books, New York)
25. E.W. Kolb, M.S. Turner, *The Early Universe* (Addison-Wesley, Reading, 1990)
26. S. Weinberg, *Cosmology* (Oxford University Press, Oxford, 2008)
27. B. Ryden, *Introduction to Cosmology* (Cambridge University Press, Cambridge, 2017)
28. S. Perlmutter et al., Measurements of Ω and Λ from 42 high-redshift supernovae. Astrophys. J. **517**, 565 (1999)
29. A.G. Riess et al., Type Ia supernova discoveries at $z > 1$ from the hubble space telescope: evidence for past deceleration and constraints on dark energy evolution. Astrophys. J. **607**, 665 (2004)
30. Y. Akrami et al., [Planck Collaboration] Planck 2018 results. I. Overview and the cosmological legacy of planck (2018), arXiv:1807.06205 [astro-ph.CO]
31. A. Gabrielli, M. Joyce, F.S. Labini, Glass-like universe: real-space correlation properties of standard cosmological models. Phys. Rev. D **65**, 083523 (2002)
32. L. Anderson et al., [BOSS Collaboration] The clustering of galaxies in the SDSS-III Baryon oscillation spectroscopic survey: Baryon acoustic oscillations in the data releases 10 and 11 galaxy samples. Mon. Not. R. Astron. Soc. **441**, 24 (2014)
33. W. Hu, N. Sugiyama, J. Silk, The physics of the microwave background anisotropies. Nature **386**, 37 (1997)
34. S. Coleman, *Aspects of Symmetry* (Cambridge University Press, Cambridge, 1985)
35. A.D. Sakharov, The initial stage of an expanding universe and the appearance of a nonuniform distribution of matter. Sov. Phys. JETP **22**, 241 (1966); [J. Exptl. Theoret. Phys. (U.S.S.R.) **49**, 345 (1965)]
36. V.F. Mukhanov, G.V. Chibisov, Quantum fluctuations and a nonsingular universe. JETP Lett. **33**, 532 (1981); [Pisma Zh. Eksp. Teor. Fiz. **33**, 549 (1981)]
37. G. Hinshaw et al., Nine-Year Wilkinson Microwave Anisotropy Probe (WMAP) observations: cosmological parameter results. Astrophys. J. Suppl. Ser. **208**, 19 (2013)
38. A.H. Guth, Quantum fluctuations in cosmology and how they lead to a multiverse (2013), arXiv:1312.7340 [hep-th]
39. M. Abramowitz, I.A. Stegun (eds.), *Handbook of Mathematical Functions*, 9th edn. (Dover, New York, 1972)
40. I.S. Gradshteyn, I.W. Ryzhik, *Tables of Integrals, Series, and Products*, 6th edn. (Academic, New York, 2000)
41. V. Mukhanov, *Physical Foundations of Cosmology* (Cambridge University Press, Cambridge, 2005)
42. R.H. Cyburt, B.D. Fields, K.A. Olive, T.-H. Yeh, Big bang nucleosynthesis: present status. Rev. Mod. Phys. **88**, 015004–1 (2016)
43. M. Tanabashi et al., (Particle Data Group) Review of particle physics. Phys. Rev. D **98**, 030001 (2018)
44. G. Bertone, D. Hooper, A history of dark matter. Rev. Mod. Phys. **90**, 45002 (2018)
45. G. Squires et al., The dark matter, gas, and galaxy distributions in Abell 2218: a weak gravitational lensing and X-ray analysis. Astrophys. J. **461**, 572 (1995)
46. M. Persic, P. Salucci, F. Stel, The universal rotation curve of spiral Galaxies - I. The dark matter connection. Mon. Not. R. Astron. Soc. **281**, 27 (1996)
47. M. Honma, Y. Sofue, On the Keplarian rotation curves of galaxies. Pub. Astron. Soc. Jpn. **49**, 539 (1997)
48. J. Binney, S. Tremaine, *Galactic Dynamics* (Princeton University Press, Princeton, 2008)
49. H. Mo, F. van den Bosch, S. White, *Galaxy Formation and Evolution* (Cambridge University Press, Cambridge, 2010)

50. J.F. Navarro, C.S. Frenk, S.D.M. White, D.M. Simon, The structure of cold dark matter halos. Astrophys. J. **462**, 563 (1996)
51. A. Burkert, The structure of dark matter halos in Dwarf galaxies. Astrophys. J. Lett. **447**, L25 (1995)
52. F. Nesti, P. Salucci, The dark matter halo of the Milky Way. AD **2013**, (2013). arXiv:1304.5127
53. R. Bottema, J.L.G. Pestana, B. Rothberg, R.H. Sanders, MOND rotation curves for spiral galaxies with cepheid-based distances. Astron. Astrophys. **393**, 453 (2002)
54. Y. Sofue, Rotation and mass in the milky way and spiral galaxies. Pub. Astron. Soc. Jpn. **69**, R1 (2017)
55. F.L. Zhi, L.S. Xian, T. Kiang, *Creation of the Universe* (World Scientific, Singapore, 1989)
56. D.A. McQuarrie, *Statistical Mechanics* (University Science Books, Sausalito, 2000)
57. F. Couchot et al., Cosmological constraints on the neutrino mass including systematic uncertainties (2017), arXiv:1703.10829 [astro-ph.CO]
58. W. Grandy, *Foundations of Statistical Mechanics* (Springer, New York, 1987)
59. C. G. Böhmer, T. Harko, *Can dark matter be a Bose-Einstein condensate?* (2007), arXiv:0705.4158v4 [astro-ph]
60. V.T. Toth, Self-gravitating Bose-Einstein condensates and the Thomas-Fermi approximation (2016), arXiv:1402.0600 [gr-qc]
61. A. Suárez, V.H. Robles, T. Matos, A review on the scalar field/Bose-Einstein condensate dark matter model (2013), arXiv:1302.0903 [astro-ph.CO]

Index

© Springer Nature Switzerland AG 2020
R. Hentschke and C. Hölbling, *A Short Course in General Relativity
and Cosmology*, Undergraduate Lecture Notes in Physics,
https://doi.org/10.1007/978-3-030-46384-7